CAMBRIDGE STUDIES IN ADVANCED MATHEMATICS 160

REPRESENTATIONS OF THE INFINITE SYMMETRIC GROUP

Representation theory of big groups is an important and quickly developing part of modern mathematics, giving rise to a variety of important applications in probability and mathematical physics. This book provides the first concise and self-contained introduction to the theory on the simplest, yet very nontrivial, example of the infinite symmetric group, focusing on its deep connections to probability, mathematical physics, and algebraic combinatorics.

Following a discussion about the classical Thoma theorem, which describes the characters of the infinite symmetric group, the authors describe explicit constructions of an important class of representations, including both the irreducible and generalized regular ones. Complete with detailed proofs, as well as numerous examples and exercises which help to summarize recent developments in the field, this book will enable graduates to enhance their understanding of the topic while also aiding lecturers and researchers in related areas.

Alexei Borodin is a Professor of Mathematics at the Massachusetts Institute of Technology. He also holds the position of Principal Researcher at the Institute for Information Transmission Problems of the Russian Academy of Sciences, Moscow.

Grigori Olshanski is a Principal Researcher in the Section of Algebra and Number Theory at the Institute for Information Transmission Problems of the Russian Academy of Sciences, Moscow. He also holds the position of Dobrushin Professor at the National Research University Higher School of Economics, Moscow.

CAMBRIDGE STUDIES IN ADVANCED MATHEMATICS

All the titles listed below can be obtained from good booksellers or from Cambridge University Press. For a complete series listing visit: www.cambridge.org/mathematics.

Already published

124 K. Lux & H. Pahlings *Representations of groups*
125 K. S. Kedlaya *p-adic differential equations*
126 R. Beals & R. Wong *Special functions*
127 E. de Faria & W. de Melo *Mathematical aspects of quantum field theory*
128 A. Terras *Zeta functions of graphs*
129 D. Goldfeld & J. Hundley *Automorphic representations and L-functions for the general linear group, I*
130 D. Goldfeld & J. Hundley *Automorphic representations and L-functions for the general linear group, II*
131 D. A. Craven *The theory of fusion systems*
132 J. Väänänen *Models and games*
133 G. Malle & D. Testerman *Linear algebraic groups and finite groups of Lie type*
134 P. Li *Geometric analysis*
135 F. Maggi *Sets of finite perimeter and geometric variational problems*
136 M. Brodmann & R. Y. Sharp *Local cohomology (2nd Edition)*
137 C. Muscalu & W. Schlag *Classical and multilinear harmonic analysis, I*
138 C. Muscalu & W. Schlag *Classical and multilinear harmonic analysis, II*
139 B. Helffer *Spectral theory and its applications*
140 R. Pemantle & M. C. Wilson *Analytic combinatorics in several variables*
141 B. Branner & N. Fagella *Quasiconformal surgery in holomorphic dynamics*
142 R. M. Dudley *Uniform central limit theorems (2nd Edition)*
143 T. Leinster *Basic category theory*
144 I. Arzhantsev, U. Derenthal, J. Hausen & A. Laface *Cox rings*
145 M. Viana *Lectures on Lyapunov exponents*
146 J.-H. Evertse & K. Győry *Unit equations in Diophantine number theory*
147 A. Prasad *Representation theory*
148 S. R. Garcia, J. Mashreghi & W. T. Ross *Introduction to model spaces and their operators*
149 C. Godsil & K. Meagher *Erdős–Ko–Rado theorems: Algebraic approaches*
150 P. Mattila *Fourier analysis and Hausdorff dimension*
151 M. Viana & K. Oliveira *Foundations of ergodic theory*
152 V. I. Paulsen & M. Raghupathi *An introduction to the theory of reproducing kernel Hilbert spaces*
153 R. Beals & R. Wong *Special functions and orthogonal polynomials*
154 V. Jurdjevic *Optimal control and geometry: Integrable systems*
155 G. Pisier *Martingales in Banach spaces*
156 C. T. C. Wall *Differential topology*
157 J. C. Robinson, J. L. Rodrigo & W. Sadowski *The three-dimensional Navier–Stokes equations*
158 D. Huybrechts *Lectures on K3 surfaces*
159 H. Matsumoto & S. Taniguchi *Stochastic analysis*
160 A. Borodin & G. Olshanski *Representations of the infinite symmetric group*
161 P. Webb *A course in finite group representation theory*
162 C. J. Bishop & Y. Peres *Fractals in probability and analysis*
163 A. Bovier *Gaussian processes on trees*

Representations of the Infinite Symmetric Group

ALEXEI BORODIN
Massachusetts Institute of Technology
and
Institute for Information Transmission Problems of the Russian Academy of Sciences

GRIGORI OLSHANSKI
Institute for Information Transmission Problems of the Russian Academy of Sciences
and
National Research University Higher School of Economics, Moscow

CAMBRIDGE
UNIVERSITY PRESS

Shaftesbury Road, Cambridge CB2 8EA, United Kingdom

One Liberty Plaza, 20th Floor, New York, NY 10006, USA

477 Williamstown Road, Port Melbourne, VIC 3207, Australia

314–321, 3rd Floor, Plot 3, Splendor Forum, Jasola District Centre, New Delhi – 110025, India

103 Penang Road, #05–06/07, Visioncrest Commercial, Singapore 238467

Cambridge University Press is part of Cambridge University Press & Assessment, a department of the University of Cambridge.

We share the University's mission to contribute to society through the pursuit of education, learning and research at the highest international levels of excellence.

www.cambridge.org
Information on this title: www.cambridge.org/9781107175556

DOI: 10.1017/9781316798577

First published 2017

A catalogue record for this publication is available from the British Library

Library of Congress Cataloging-in-Publication data
Names: Borodin, Alexei. | Olshanskiĭ, G. I. (Grigori I.)
Title: Representations of the infinite symmetric group / Alexei Borodin, Massachusetts Institute of Technology, Grigori Olshanskiĭ, Institute for Information Transmission Problems of the Russian Academy of Sciences and National Research University Higher School of Economics, Moscow.
Description: Cambridge : Cambridge University Press, [2017] | Series: Cambridge studies in advanced mathematics ; 160 | Includes bibliographical references and index.
Identifiers: LCCN 2016025925 | ISBN 9781107175556
Subjects: LCSH: Hopf algebras. | Algebraic topology. | Representations of groups. | Symmetry groups.
Classification: LCC QA613.8 .B67 2017 | DDC 515/.22–dc23
LC record available at https://lccn.loc.gov/2016025925

ISBN 978-1-107-17555-6 Hardback

Contents

Introduction

As is well known, representation theory began with the work of G. Frobenius, who invented the notion of a character of a noncommutative finite group and solved a highly nontrivial problem of describing irreducible characters of symmetric groups $S(n)$ (see his paper [41] of 1900, and also Curtis [30]). Twenty-five years later, H. Weyl computed the irreducible characters of the compact classical Lie groups $U(N)$, $SO(N)$, $Sp(N)$ (see his famous book [133] and references therein). These results of Frobenius and Weyl form the basis of the whole representation theory of groups, and in one way or another they usually appear in any introductory course on finite-dimensional group representations (e.g., Fulton and Harris [42], Goodman and Wallach [52], Simon [111], Zhelobenko [136]).

It is a remarkable fact that character theory can be built for *infinite-dimensional analogs* of symmetric and classical groups if one suitably modifies the notion of the character. This was independently discovered by E. Thoma in the 1960s [117] for the infinite symmetric group $S(\infty)$ and by D. Voiculescu in the 1960s [130], [131] for infinite-dimensional classical groups $U(\infty)$, $SO(\infty)$, $Sp(\infty)$. It turned out that, for all these groups, the so-called extreme characters (analogs of the irreducible characters) depended on countably many continuous parameters, and in the two cases, i.e., for $S(\infty)$ and infinite-dimensional classical groups, the formulas looked very similar.

In spite of all the beauty of Thoma's and Voiculescu's results, they looked too unusual and even exotic, and were largely away from the principal routes of representation theory that formed the mainstream in the 1960s and 70s. It took time to appreciate their depth and realize what kind of mathematics lies behind them. Thoma's and Voiculescu's original motivation came from the theory of von Neumann factors and operator algebras. Nowadays we can point out some connections with other areas of mathematics. First of all, those are:

(a) combinatorics of symmetric functions and multivariate special functions of hypergeometric type;

and

(b) probabilistic models of mathematical physics: random matrices, determinantal point processes, random tiling models, Markov processes of infinitely many interacting particles. Let us emphasize that such connections to probability theory and mathematical physics are new; previously known ones were of a different kind (see, e.g., P. Diaconis' book [31]).

Pioneering work of A.M. Vershik and S.V. Kerov (see [63], [121], [122], [123], [124], [125], [126]) played a key role in bringing forward the combinatorial and probabilistic aspects of the representation theory of $S(\infty)$ and infinite-dimensional classical groups. We will say more about their work below.

—

The goal of this book is to provide a detailed introduction to the representation theory of the infinite symmetric group that would be accessible to graduate and advanced undergraduate students. The amount of material that would be required for the reader to know in advance is rather modest: some familiarity with representation theory of finite groups and basics of functional analysis (measure theory, Stone–Weierstrass' theorem, Choquet's theorem on extreme points of a compact set, Hilbert spaces) would suffice. Theory of symmetric functions plays an important role in our approach, and while some previous exposure to it would be useful, we also provide all the necessary background along the way.

We aimed at writing a relatively short, simple, and self-contained book and did not try to include everything people know about representations of $S(\infty)$. We also do not touch upon representations of infinite-dimensional classical groups – while being analogous, that theory is somewhat more involved. The infinite symmetric group can be viewed as a "toy model" for infinite-dimensional classical groups. We feel that it makes sense to begin the exposition with $S(\infty)$, in parallel to how the subject of representation theory historically and logically started from the finite symmetric groups $S(n)$ that form the simplest and most natural family of finite noncommutative groups.

Knowledge of the material in this book should be sufficient for understanding research papers on representations of $S(\infty)$ and infinite-dimensional groups as well as their applications. Speaking of applications, we first of all mean the class of probabilistic models of mathematical physics where representation theoretic ideas turned out to be remarkably successful. We hope that probabilists and mathematical physicists interested in representation theoretic mechanisms behind such applications would find the book useful.

—

We have not defined our main object of study yet; let us do that now. One could give different (meaningful) definitions of the infinite symmetric group. In this book we define $S(\infty)$ as the group of all finite permutations of the set $\mathbb{Z}_{>0} := \{1, 2, 3, \ldots\}$, where the condition of a permutation $s : \mathbb{Z}_{>0} \to \mathbb{Z}_{>0}$ being finite means that $s(j) \neq j$ for finitely many $j \in \mathbb{Z}_{>0}$. Equivalently, $S(\infty)$ can be defined as the inductive limit (or simply the union) of the infinite chain $S(1) \subset S(2) \subset S(3) \subset \cdots$ of growing finite symmetric groups. The group $S(\infty)$ is countable, and it is reasonable to view it as a natural (yet not canonical) infinite analog of the finite symmetric groups $S(n)$. Infinite-dimensional classical groups are defined in a similar fashion as inductive limits of finite-dimensional classical Lie groups.

The first part of the book deals with characters of $S(\infty)$. Similarly to the case of finite groups, a substantial part of representation theory can be built in the language of characters without even mentioning actual representations. Many applications would just require operating with characters. We believe, however, that ignoring representations behind the characters takes away an essential part of the subject and may eventually negatively influence future developments. For that reason, in the second part we turn to unitary representations.

Let us now describe the content of the book in a little more detail.

—

The first part of the book is devoted to Thoma's theorem and related topics. As was mentioned above, Thoma's theorem is an analog of the classical Frobenius theorem on irreducible characters of the finite symmetric groups. The word "analog" should be taken with a pinch of salt here. The point is that $S(\infty)$ does not have conventional irreducible characters (except for two trivial examples), and the notion needs to be revised. Here is a definition given by Thoma that we use:

(a) A *character* of a given group K is a function $\chi : K \to \mathbb{C}$ that is positive definite, constant on conjugacy classes, and normalized to take value 1 at the unity of the group. (For topological groups one additionally assumes that the function χ is continuous.)

(b) An *extreme character* is an extreme point of the set of all characters. (This makes sense as the characters, as defined in (a), form a convex set.)

In the case where the group K is finite or compact, the set of all characters (in the sense of the above definition) is a simplex whose vertices are the extreme characters. Those are exactly the *normalized* irreducible characters, i.e., functions of the form $\chi^\pi(g)/\chi^\pi(e)$, where π denotes an arbitrary irreducible representation (it is always finite-dimensional), $\chi^\pi(g) = \operatorname{Tr}\pi(g)$ is the trace of the operator $\pi(g)$ corresponding to a group element $g \in K$, and the denominator $\chi^\pi(e)$ (the value of χ^π at the unity of the group) coincides with the dimension of the representation. For $S(\infty)$ and its relatives the numerator and denominator in

$$\frac{\chi^\pi(g)}{\chi^\pi(e)} = \frac{\operatorname{Tr}\pi(g)}{\operatorname{Tr}\pi(e)}$$

do not make sense when considered separately, but the notion of extreme character assigns a meaning to their ratio.

According to Frobenius' theorem, irreducible characters of $S(n)$ are parameterized by Young diagrams with n boxes. According to Thoma's theorem, extreme characters of $S(\infty)$ are parameterized by points of an infinite-dimensional space Ω situated inside the infinite-dimensional unit cube; a point $\omega \in \Omega$ is a pair (α, β) of infinite sequences with entries from $[0, 1]$ such that

$$\alpha = (\alpha_1 \geq \alpha_2 \geq \cdots), \quad \beta = (\beta_1 \geq \beta_2 \geq \cdots), \quad \sum_{i=1}^{\infty} \alpha_i + \sum_{i=1}^{\infty} \beta_i \leq 1.$$

Direct comparison of the two theorems leads to a somewhat perplexing conclusion that the extreme characters of $S(\infty)$ are both simpler and more complicated than the irreducible characters of $S(n)$. They are more complicated because, instead of the finite set of Young diagrams with n boxes, one gets a countable set of continuous parameters α_i, β_i. But at the same time they are simpler because, for the extreme characters, there is an explicit elementary formula (found by Thoma), while the irreducible characters of finite symmetric groups should be viewed as special functions – there are algorithms for computing them but no explicit formulas.

Thoma's theorem can be (nontrivially) reformulated in several ways. In particular, it is equivalent to:

(1) classifying infinite upper-triangular Toeplitz matrices, all of whose minors are nonnegative (matrices with nonnegative minors are called *totally positive*);
(2) describing all multiplicative functionals on the algebra of symmetric functions that take nonnegative values on the basis of the Schur symmetric functions;
(3) describing the entrance boundary for a certain Markov chain related to the Young graph or, equivalently, describing the extreme points in a suitably defined space of Gibbs measures on paths of the Young graph.

The original proof by Thoma consisted in reduction to (1) and solving the classification problem. Apparently, Thoma did not know that the latter problem had been studied earlier by Schoenberg and his followers ([109], [1], [2]) and its solution had been finalized by Edrei [36].[1]

Interpretations (2) and (3) are due to Vershik and Kerov [122], [124]. The proof of Thoma's theorem that we give is based on (3) and it is an adaptation of an argument from the paper [64] by Kerov, Okounkov, and Olshanski, where a more general result had been proved. We do not focus on extreme characters but rather establish an isomorphism between the convex set of all characters and the convex set of all probability measures on Ω, which immediately implies Thoma's theorem. Although our approach deviates from the proof outlined by Vershik and Kerov in [122], we substantially rely on their *asymptotic method*, which, in particular, explains the nature of Thoma's parameters $\{\alpha_i, \beta_i\}$.

The solution of problem (1) given by Edrei [36] and Thoma [117] is largely analytic, while our approach to the equivalent problem (3) is in essence algebraic; we work with symmetric functions and rely on results of Okounkov, Olshanski [80], and Olshanski, Regev, and Vershik [97], [98].

—

In the second part of the book we move from characters to representations. Our exposition is based on the works of Olshanski [85] and Kerov, Olshanski, and Vershik [67].

There exist two approaches that relate characters to unitary representations. To be concrete, let us discuss extreme characters $\chi = \chi^\omega$ of $S(\infty)$, where $\omega = (\alpha, \beta) \in \Omega$ and (α, β) are Thoma's parameters of χ. The first approach gives the corresponding factor-representations Π^ω of the group $S(\infty)$ (Thoma [117]), while in the second approach one deals with irreducible representations T^ω of the "bisymmetric group" $S(\infty) \times S(\infty)$ (Olshanski [84]). Factor-representations Π^ω are analogs of the irreducible representations π^λ of the finite symmetric groups, but they are (excluding two trivial cases) not irreducible at all: the term "factor" means that these are unitary representations that generate a von Neumann factor (in our case, this is the hyperfinite factor of type II_1). The representations T^ω are exactly those irreducible representations of the bisymmetric group $S(\infty) \times S(\infty)$ that contain a nonzero vector that is invariant with respect to the subgroup diag $S(\infty)$ (the image of $S(\infty)$ under its

[1] Likewise, the extreme characters of $U(\infty)$ correspond to arbitrary (not necessarily upper triangular) totally positive Toeplitz matrices; such matrices were classified, prior to Voiculescu's work, in another paper by Edrei [37]. At the present time, the theory of total positivity became popular due to works of G. Lusztig, S. Fomin, and A. Zelevinsky, but in the sixties and seventies it was much less known.

diagonal embedding in $S(\infty) \times S(\infty)$). Such representations are called *spherical*. Similar representations for finite bisymmetric groups $S(n) \times S(n)$ have the form $\pi^\lambda \otimes \pi^\lambda$, but representations T^ω cannot be written as a (exterior) tensor product of two irreducibles. Such a phenomenon is typical for so-called *wild* (or non-type I) groups, and the infinite symmetric group is one of them.

For wild groups the space of equivalence classes of irreducible representations has pathological structure, and for that reason standard representation theoretic problem settings require modifications. There are many interesting examples of wild groups but there is no general recipe for addressing their representation theories. This highlights the remarkable fact that, for the infinite symmetric group and the infinite-dimensional classical groups, it is possible to develop a meaningful representation theory, and the purpose of the second part of the book is to give an introduction to this theory.

Returning to factor-representations Π^ω and irreducible spherical representations T^ω, let us note that they are closely related: namely, Π^ω is the restriction of T^ω to the subgroup $S(\infty) \times \{e\}$ in the bisymmetric group (that should not be confused with the diagonal subgroup diag $S(\infty)$!). Thus, to a certain extent, the choice between factor-representations of $S(\infty)$ and irreducible representations of $S(\infty) \times S(\infty)$ is a matter of taste (the theories do diverge in further developments, though). We follow the approach of Olshanski [84] and prefer to work with irreducible representations T^ω.

The existence of such representations is a simple corollary of Thoma's theorem. However, that theorem gives no information as to how such representations could be explicitly constructed. This is a typical representation theoretic situation, when it is known how to parameterize the representations but their explicit construction may be completely unobvious and very complicated.

The first realization of representations T^ω was found by Vershik and Kerov [121] (actually, they dealt with factor-representations Π^ω but from here the passage to T^ω is easy). We describe a modification of their construction that employs infinite tensor products of Hilbert spaces in the sense of von Neumann. For generic values of Thoma's parameters, when α- and β-parameters of ω are nonzero, the Hilbert spaces involved in the tensor product construction have a $\mathbb{Z}_2$-grading, i.e., they are super-spaces, and their tensor product has to be understood according to the sign rule from linear super-algebra. The super-algebra actually already appears in the first half of the book – it is present in the formula for Thoma's characters, where super-analogs of Newton power sums arise. Representation theory of the infinite symmetric group is an example of a subject where the need for the use of supersymmetric notions is dictated by the nature of the objects involved rather than by a pure wish to generalize known results to a super-setting.

Let us note that the infinite tensor product construction also gives more general, so-called admissible representations of the bisymmetric group (see Olshanski [85]), and that there is also one more realization due to Okounkov [78], [79].

Next, following the general philosophy of unitary representation theory, we proceed from the theory of irreducible representations to harmonic analysis. The problem of (noncommutative) harmonic analysis is, to a certain extent, similar to Fourier analysis or to expansion in eigenfunctions of a self-adjoint operator. One starts with a ("natural" in some sense) reducible unitary representation, and the problem consists in finding its decomposition on irreducible components. As for self-adjoint operators, the spectrum of the decomposition may be complicated, e.g., not necessarily discrete. In the case of a continuous spectrum one talks about decomposing a representation into a direct integral (rather than a direct sum).

It is a matter of discussion which representations should be considered as "natural" objects for harmonic analysis. However, each finite or compact group has one distinguished representation – the regular representation in the Hilbert space L^2 with respect to the Haar measure on the group. The action of the group is given by left (equivalently, right) shifts. It is even better to consider both left and right shifts together, i.e., the so-called *bi-regular representation* of the direct product of two copies of the group.

For compact (in particular, finite) groups, the decomposition of the bi-regular representation is well known and quite simple (Peter–Weyl's theorem). In particular, the bi-regular representation of the bisymmetric group $S(n) \times S(n)$ is a multiplicity free direct sum of irreducible spherical representations $\pi^\lambda \otimes \pi^\lambda$ that were already mentioned above. Note now that for infinite-dimensional classical groups there is no Haar measure (they are not locally compact) and, therefore, there is no (bi)regular representation.

At first glance, for $S(\infty)$ the situation is different – it is a discrete countable group that carries a Haar measure (which is simply the counting measure), and its bi-regular representation makes perfect sense. However, it ends up being irreducible and thus useless for harmonic analysis.

This dead end turns out to be illusory, and we explain how it can be overcome. The essence of the problem is in the fact that the discrete group $S(\infty)$ is too small to carry a suitable measure with respect to which we would like to build the L^2 space. The way out is in constructing a compactification $\mathfrak{S} \supset S(\infty)$, called the *space of virtual permutations*, that serves as the support of the measure (Kerov, Olshanski, and Vershik [66]).[2] The space $\mathfrak{S}$

[2] The construction of the space $\mathfrak{S}$ was inspired by Pickrell's paper [102]. A close (but not identical) construction is that of *Chinese Restaurant Process*; see Pitman [104].

does not have a group structure, but $S(\infty) \times S(\infty)$ acts on it, and there is a (unique) invariant measure μ on $\mathfrak{S}$ which is finite, as opposed to the counting measure on $S(\infty)$. It is that measure that should be viewed as the correct analog of the Haar measure. Furthermore, the measure μ is just a representative of a whole family of probability measures with good transformation properties that are equally suitable for constructing representations.[3] One thus obtains a whole family $\{T_z\}$ of *generalized bi-regular representations* that depend on a parameter $z \in \mathbb{C}$. The problem of harmonic analysis in our understanding is the problem of decomposing these representations on irreducibles.

We prove that each T_z decomposes on irreducible spherical representations T^ω, and that the decomposition spectrum is simple. We also prove that the spectral measures that govern the decomposition of T_z are mutually singular. This result is important as it implies that the representations T_z are pairwise disjoint, and thus the parameter z is not fictitious.

Investigating the spectral measures goes beyond the goals of this book. It turns out that their structure substantially depends on whether parameter z is an integer or not. These two cases are studied separately and using different means in Kerov, Olshanski, and Vershik [67], and in Borodin and Olshanski [9], respectively. The case of non-integral z is especially interesting as, in the course of its study, one discovers novel models of determinantal random point processes and close connections to random matrix theory.

We conclude with a (short and incomplete) guide to the literature encompassing other aspects of the theory and its further development:

- *A few expository papers* (unfortunately, already rather old): Borodin and Olshanski [8], [12], [14], Olshanski [92].
- *Asymptotic approach to characters of infinite-dimensional classical groups*: Vershik and Kerov [123], Okounkov and Olshanski [81], [82], Borodin and Olshanski [22].
- *Irreducible unitary representations of infinite-dimensional classical groups*: Olshanski [84], [87], [86], Pickrell [103].
- *Quasiinvariant measures for infinite-dimensional classical groups*: Pickrell [102], Neretin [75].
- *Harmonic analysis on $U(\infty)$*: Olshanski [91], Borodin and Olshanski [13], Gorin [54], Osinenko [99].

[3] These measures, called *Ewens measures*, are very interesting in their own right. They are closely related to the *Ewens sampling formula* that is widely used in the literature on mathematical models of population genetics; see, e.g., the survey paper by Ewens and Tavaré [38].

- *The Plancherel measure on partitions*: Kerov [60], [61], [63], Baik, Deift, and Johansson [4], Johansson [58], Borodin, Okounkov, and Olshanski [7], Strahov [115].
- *Other measures on partitions of representation-theoretic origin and their generalizations*: Borodin and Olshanski [9], [10], [15], [16], [18], [24], Borodin, Olshanski and Strahov, [25], Olshanski [90], [93], [96], Strahov [116].
- *Models of Markov dynamics of representation-theoretic origin*: Borodin and Olshanski [17], [19], [20], [21], Olshanski [94], [95].

Acknowledgments. We are grateful to Cesar Cuenca, Vadim Gorin, Leonid Petrov, and the anonymous referee whose remarks helped us to fix typos and improve the exposition. The work of the second-named author (G.O.) was partially supported by the Simons Foundation (the Simons–IUM fellowship).

PART ONE

SYMMETRIC FUNCTIONS AND THOMA'S THEOREM

1

Preliminary Facts From Representation Theory of Finite Symmetric Groups

A *partition* of a natural number n is a weakly decreasing sequence of non-negative integers which add up to n:

$$\lambda = (\lambda_1 \geq \lambda_2 \geq \cdots \geq 0), \qquad \lambda_1 + \lambda_2 + \cdots = n.$$

Partitions are often pictured by *Young diagrams* (also called *Ferrers diagrams*): the diagram corresponding to a given partition λ of n is the left-justified collection of n boxes containing λ_i boxes in the ith row, counting from top to bottom (Sagan [107, Definition 2.1.1]). In what follows, we identify a partition and the corresponding Young diagram. We denote the set of all partitions of n by $\mathbb{Y}_n$, and we agree that $\mathbb{Y}_0$ consists of a single element – the zero partition or the empty diagram $\varnothing$.

Given $\lambda \in \mathbb{Y}_n$, we write $|\lambda| = n$ and denote by $d = d(\lambda)$ the number of diagonal boxes in λ. We also use the *Frobenius notation* (Macdonald [72]):

$$\lambda = (p_1, \ldots, p_d \mid q_1, \ldots, q_d).$$

Here, $p_i = \lambda_i - i$ is the number of boxes in the ith row of λ to the right from the ith diagonal box; likewise, $q_i = \lambda'_i - i$ is the number of boxes in the ith column of λ below the ith diagonal box (λ' stands for the *transposed diagram*).

For instance, if $\lambda = (6, 5, 2, 0, \ldots)$, then $d = 2$, $\lambda' = (3, 3, 2, 2, 2, 1)$, and $\lambda = (5, 3 \mid 2, 1)$ in the Frobenius notation.

Note that

$$p_1 > \cdots > p_d \geq 0, \qquad q_1 > \cdots > q_d \geq 0, \qquad \sum_{i=1}^{d} (p_i + q_i + 1) = |\lambda|.$$

The numbers p_i and q_i are called the *Frobenius coordinates* of the diagram λ. Obviously, transposition of λ amounts to switching its Frobenius coordinates: $p_i \leftrightarrow q_i$.

We denote by $S(n)$ the group of permutations of the set $\{1, 2 \ldots, n\}$.

Proposition 1.1 *The conjugacy classes of the group $S(n)$ are parameterized by the elements of $\mathbb{Y}_n$. More precisely, $\lambda \in \mathbb{Y}_n$ corresponds to the class consisting of permutations which can be represented as products of disjoint cycles of lengths $\lambda_1, \lambda_2, \dots$*

This fact is easy to prove (see, e.g., Sagan [107, Section 1.1] for details). If a permutation s belongs to the conjugacy class parameterized by λ then we also say that s has the *cycle structure* λ.

Proposition 1.2 *The complex irreducible representations of the group $S(n)$ are also parameterized by $\mathbb{Y}_n$. (They are all finite-dimensional and unitary, as for any finite group.)*

The correspondence is much more involved than that for the conjugacy classes; see, e.g., Sagan [107, Section 2]. We denote by π_λ the representation corresponding to λ, and we set $\dim \lambda := \dim \pi_\lambda$.

To any box $\square$ of a Young diagram λ we assign a number $h(\square)$ called the *hook length* of the box: if $\square$ is located in row i and column j then

$$h(\square) = \lambda_i - i + \lambda'_j - j + 1.$$

We denote by $\ell(\lambda)$ the number of nonzero parts of a partition λ.

Here are four useful formulas for $\dim \lambda$.

Proposition 1.3 *Let λ be a partition, and let N be any integer $\geq \ell(\lambda)$. Set $d = d(\lambda)$. Then*

$$\frac{\dim \lambda}{|\lambda|!} = \prod_{\square \in \lambda} \frac{1}{h(\square)} \tag{1.1}$$

$$= \det \left[\frac{1}{(\lambda_i - i + j)!} \right]_{i,j=1}^{N} \tag{1.2}$$

$$= \frac{\prod_{1 \leq i < j \leq N} (\lambda_i - i - \lambda_j + j)}{\prod_{i=1}^{N} (\lambda_i + N - i)!} \tag{1.3}$$

$$= \frac{1}{\prod_{i=1}^{d} p_i! q_i!} \cdot \frac{\prod_{1 \leq i < j \leq d} (p_i - p_j)(q_i - q_j)}{\prod_{i,j=1}^{d} (p_i + q_j + 1)} . \tag{1.4}$$

Equality (1.1) is called the *hook formula*, and (1.2) is called the *Frobenius formula*; see, e.g., Sagan [107, Theorems 3.10.2 and 3.11.1]. Hints of how to prove the equivalence of all four formulas are given in the exercises to this section.

We will use the notation $\mu \nearrow \lambda$ or $\lambda \searrow \mu$ to indicate that diagram λ is obtained from diagram μ by appending a box:

$$|\lambda| = |\mu| + 1, \qquad \lambda_i = \mu_i + 1 \quad \text{for exactly one } i.$$

Given a group G, its subgroup H, and a representation π of G, we denote by $\pi|_H$ or by $\mathrm{Res}_H^G \pi$ the restriction of π to H. If G is finite, and ρ is a representation of H, we denote by $\mathrm{Ind}_H^G \rho$ the representation of G induced by ρ.

For every $n = 1, 2, \ldots$, the embedding $\{1, \ldots, n\} \subset \{1, \ldots, n+1\}$ induces a natural embedding of $S(n)$ into $S(n+1)$: namely, the $S(n)$ is identified with the stabilizer of the point $n+1$ in $S(n+1)$.

Proposition 1.4 (Branching relations) (i) *For any $n = 2, 3, \ldots$ and $\lambda \in \mathbb{Y}_n$*

$$\mathrm{Res}_{S(n-1)}^{S(n)} \pi_\lambda = \bigoplus_{\mu \in \mathbb{Y}_{n-1}:\, \mu \nearrow \lambda} \pi_\mu.$$

(ii) *Equivalently, for any $n = 1, 2, \ldots$ and $\lambda \in \mathbb{Y}_n$*

$$\mathrm{Ind}_{S(n)}^{S(n+1)} \pi_\lambda = \bigoplus_{\nu \in \mathbb{Y}_{n+1}:\, \nu \searrow \lambda} \pi_\nu.$$

The equivalence of these statements follows from the *Frobenius reciprocity*; see, e.g., Sagan [107, Theorem 1.12.6]. The first statement is often called the *Young branching rule* (Sagan [107, Theorem 2.8.3]).

The branching relations immediately imply that for any partition λ,

$$\dim \lambda = \sum_{\mu:\, \mu \nearrow \lambda} \dim \mu, \qquad \dim \lambda = \frac{1}{n+1} \sum_{\nu:\, \nu \searrow \lambda} \dim \nu. \tag{1.5}$$

Iterating the first relation we see that for every $\lambda \in \mathbb{Y}_n$, $\dim \lambda$ is equal to the number of sequences

$$\varnothing \nearrow \lambda(1) \nearrow \lambda(2) \nearrow \cdots \nearrow \lambda(n) = \lambda, \qquad \lambda(i) \in \mathbb{Y}_i.$$

These sequences can be encoded by *standard Young tableaux of shape* λ. Such a tableau is, by definition, a filling of all boxes of the Young diagram λ with numbers from 1 to n in such a way that the numbers increase from left to right and from top to bottom.

If G is a finite group and π is a finite-dimensional (complex) representation of G, then the term *character* of π usually refers to the function on G which takes value $\mathrm{Tr}(\pi(g))$ at $g \in G$. We will denote this function by χ^π.

A function φ on a group G (not necessarily finite) is called *positive definite* if for any $k = 1, 2, \ldots$, and $g_1, \ldots, g_k \in G$ the matrix $[\varphi(g_j^{-1} g_i)]_{i,j=1}^k$ is Hermitian and positive definite. Equivalently, for any $g \in G$ we have

$\varphi(g^{-1}) = \overline{\varphi(g)}$, and for any number $k = 1, 2, \ldots$ and any k-tuples $z_1, \ldots, z_k \in \mathbb{C}$, $g_1, \ldots, g_k \in G$ we have

$$\sum_{i,j=1}^{k} z_i \overline{z_j} \varphi(g_j^{-1} g_i) \geq 0.$$

A function χ on a group G is called *central* (another name is *class function*) if it is constant on conjugacy classes. Equivalently, for any $g, h \in G$ one has $\chi(gh) = \chi(hg)$.

Proposition 1.5 *Characters of complex finite-dimensional representations of finite groups have the following properties:*

- *Characters are central functions on G.*
- *The value of a character at the unit element of the group is equal to the dimension of the corresponding representation.*
- *Any character χ is a positive definite function on the group G.*
- *A character defines the corresponding representation uniquely, i.e. non-equivalent representations have unequal characters.*
- *Characters of the irreducible representations of G form an orthonormal basis in the space of all central functions on G equipped with the inner product*

$$\langle \phi, \psi \rangle = \frac{1}{|G|} \sum_{g \in G} \phi(g) \overline{\psi(g)}.$$

The first three statements are easy to prove, and the proofs of the last two can be found, e.g., in Sagan [107, Section 1.8 and 1.9] (or in any textbook on representations of finite groups).

Characters are extremely useful in studying representations of finite groups. However, for more complicated groups (like the infinite symmetric group $S(\infty)$ that we will define a little later), the representations become infinite-dimensional, and it becomes hard to define the characters using the trace. Thus, it is useful to have an independent characterization of characters.

We denote by $\widehat{G}$ the set of (equivalency classes of) irreducible representations of a finite group G.

Proposition 1.6 *For a finite group G, a function $\varphi : G \to \mathbb{C}$ is central, positive definite, and takes value 1 at the unit element of the group if and only if it is a convex combination of normalized characters of the irreducible representations of G:*

$$\varphi = \sum_{\pi \in \widehat{G}} c_\pi \frac{\chi^\pi}{\dim \pi}, \qquad c_\pi \geq 0, \qquad \sum_{\pi \in \widehat{G}} c_\pi = 1.$$

That is, all such functions form a simplex of dimension $|\widehat{G}| - 1$ whose vertices are the normalized characters of the irreducible representations of G.

Proof The "if" part is fairly obvious; it also follows from the statements of Proposition 1.5 above. Let us prove the "only if" implication.

Consider the space $\mathbb{C}[G]$ of complex-valued functions on G. Define a multiplication operation (called *convolution*) on $\mathbb{C}[G]$ by

$$(\varphi * \psi)(g) = \frac{1}{|G|} \sum_{h \in G} \varphi(gh^{-1})\psi(h).$$

We also define an involution on $\mathbb{C}[G]$ by $\varphi^*(g) := \overline{\varphi(g^{-1})}$. The space $\mathbb{C}[G]$ equipped with these two operations is called the *group algebra* of G.

The orthogonality relations for irreducible characters of finite groups and the fact that representations of finite groups are unitarizable imply

$$(\chi^\pi)^* = \chi^\pi, \qquad \chi^{\pi_1} * \chi^{\pi_2} = \begin{cases} \dfrac{\chi^{\pi_1}}{\dim \pi_1}, & \pi_1 = \pi_2, \\ 0, & \text{otherwise}, \end{cases} \qquad \pi, \pi_1, \pi_2 \in \widehat{G}.$$

If φ is a positive definite function on G and $\psi \in \mathbb{C}[G]$ is arbitrary, one readily checks that $\psi * \varphi * \psi^* \in \mathbb{C}[G]$ is also positive definite.

If φ is a central function then the last statement of Proposition 1.5 implies that $\varphi = \sum_{\pi \in \widehat{G}} a_\pi \chi^\pi$. Evaluating $\chi^\pi * \varphi * (\chi^\pi)^*$ for $\pi \in \widehat{G}$ we see that if φ is also positive definite then $a_\pi \geq 0$. The fact that $\sum_\pi a_\pi \dim \pi = 1$ follows from the evaluation of both sides at the unit element. □

Proposition 1.6 motivates the following modification of the conventional definition of a character. This is the definition that we will use.

Definition 1.7 Let G be an arbitrary group. By a *character* of G we mean a function $\chi : G \to \mathbb{C}$ which is central, positive definite (the explanation is in Proposition 1.5 above), and takes value 1 at the unit element.

If G is finite and π is a finite-dimensional representation of G then

$$g \mapsto \operatorname{Tr}(\pi(g))/\dim \pi$$

is a character of G in the sense of this new definition. Let us emphasize once more – the purpose of Definition 1.7 is to extend the notion of the character to groups with infinite-dimensional representations.

Proposition 1.6 may be viewed as an analog of the classical *Bochner theorem* (see, e.g., Reed and Simon [106, Theorem IX.9]), which says that the Fourier transform on $\mathbb{R}$ establishes a bijection between continuous positive definite functions on $\mathbb{R}$ which take value 1 at the origin and probability measures on $\mathbb{R}$.

The role of a probability measure on $\mathbb{R}$ is played by the collection $\{c_\pi\}_{\pi\in\widehat{G}}$, which can be viewed as a probability measure on $\widehat{G}$, and the normalized traces $\mathrm{Tr}(\pi(\cdot))/\dim\pi$ are the analogs of the 1-dimensional continuous unitary representations $x \mapsto e^{ipx}$ of the Abelian topological group $\mathbb{R}$ (which are exactly all irreducible continuous unitary representations of $\mathbb{R}$.)

Here are three examples of characters of $S(n)$.

(1) The trivial character $\chi \equiv 1$ corresponds to the trivial representation of $S(n)$ and $\lambda = (n) \in \mathbb{Y}_n$.
(2) The sign character $\chi : \sigma \mapsto \operatorname{sgn}\sigma \in \{\pm 1\}$ corresponds to the one-dimensional sign representation of $S(n)$ and $\lambda = (1, 1, \ldots, 1) \in \mathbb{Y}_n$.
(3) The delta-function at the unit element of $S(n)$ (or any other finite group) is the character (= normalized trace) of the *regular representation* of $S(n)$ in the group algebra $\mathbb{C}[S(n)]$; the action on functions is given by shifts of the argument: $(\sigma \cdot \varphi)(\tau) = \varphi(\sigma^{-1}\tau)$.

For an arbitrary finite group G, the delta-function at the unit element is expanded on irreducible characters in the following way:

$$\delta_e = \frac{1}{|G|} \sum_{\pi\in\widehat{G}} \dim\pi\ \chi^\pi .$$

Rewriting this as

$$\delta_e = \sum_{\pi\in\widehat{G}} \frac{\dim^2\pi}{|G|} \frac{\chi^\pi}{\dim\pi}$$

we get a representation of the delta-function as a convex combination of normalized irreducible characters. Indeed, the coefficients of the latter expansion are positive and sum up to 1, as it is seen by evaluating both sides at e. Another way to see this is to apply the *Burnside identity* (Sagan [107, Proposition 1.10.1, item 2]):

$$\sum_{\pi\in\widehat{G}} (\dim\pi)^2 = |G|.$$

It follows that

$$\left\{ \frac{(\dim\pi)^2}{|G|} \,\middle|\, \pi \in \widehat{G} \right\}$$

is a probability measure on $\widehat{G}$. In particular, in the symmetric group case, $G = S(n)$, we get a probability measure on $\widehat{S(n)} = \mathbb{Y}_n$,

$$\left\{ \frac{(\dim\lambda)^2}{n!} \,\middle|\, \lambda \in \mathbb{Y}_n \right\},$$

which is called the *Plancherel measure* on partitions.

1.1 Exercises

Exercise 1.1 Prove that the expressions given by (1.2) and (1.3) coincide. (Hint: use the fact that for any sequence of monic polynomials $p_n(x) = x^n + \cdots$ the determinant $\det[p_{i-1}(x_j)]_{i,j=1}^N$ coincides (up to a sign) with the Vandermonde determinant in the variables $x_1, \dots, x_N$.)

Exercise 1.2 Show that (1.3) is independent of the choice of $N \geq \ell(\lambda)$.

Exercise 1.3 Show that the hook formula (1.1), written in terms of the row coordinates $\{\lambda_i\}$, yields (1.3), while in terms of the Frobenius coordinates $\{p_i, q_i\}$ it yields (1.4).

Exercise 1.4 Prove the relations (1.5) directly by using the formula (1.3) for the dimensions. Prove that (1.5) imply the following special case of the Burnside identity:

$$\sum_{\lambda \in \mathbb{Y}_n} \dim^2 \lambda = n!.$$

Exercise 1.5 Let G be a finite group. Prove that the normalized traces $\chi^\pi / \dim \pi$ of irreducible representations $\pi \in \widehat{G}$ can be characterized as those nonzero functions $\chi : G \to \mathbb{C}$ that satisfy the relation

$$\frac{1}{|G|} \sum_{h \in G} \chi(g_1 h g_2 h^{-1}) = \chi(g_1)\chi(g_2), \qquad g_1, g_2 \in G.$$

This relation is often referred to as the *functional equation* for irreducible characters of G. Note that if G is Abelian then we just get $\chi(g_1 g_2) = \chi(g_1)\chi(g_2)$.

The functional equation provides one more way of characterizing the irreducible characters without involving the trace in the representation space. Later we will see what the functional equation looks like for the infinite symmetric group $S(\infty)$.

Exercise 1.6 Consider the action of $S(n)$ on $\mathbb{C}^n$ by permutations of coordinates. Then the orthogonal complement to the vector $(1, 1, \dots, 1)$ is an irreducible representation. The corresponding character (normalized trace) is equal to

$$\chi(\sigma) = \frac{\#\{\text{trivial cycles of } \sigma\} - 1}{n - 1}.$$

One can show that this representation corresponds to $\lambda = (n-1, 1)$.

Exercise 1.7 Prove that for $n \geq 5$ the dimension of any irreducible representation π_λ of $S(n)$, with the exception of the two one-dimensional representations, is at least $n-1$. This minimum is reached at exactly two conjugate partitions $\lambda = (n-1, 1)$ or $(2, 1, \dots, 1)$.

One can use this fact to derive that $S(\infty)$ has no finite-dimensional representations which are not direct sums of the one-dimensional ones.

Exercise 1.8 (a) Let G be a finite group such that each element $g \in G$ is conjugate to its inverse g^{-1}. Show that any character of G is real-valued.

(b) Check that the symmetric group $S(n)$ satisfies the assumption in (a) and hence its characters are real-valued. (Because of this fact, we will be dealing with real-valued functions on the groups $S(n)$.)

1.2 Notes

Almost all the material of this section is standard (the only exception is Definition 1.7). We have chosen as the main reference Sagan's book [107], but the reader may also consult other textbooks and monographs, e.g., Ceccherini-Silberstein, Scarabotti, and Tolli [29], Fulton and Harris [42], James and Kerber [57], Simon [111], Vinberg [129], Zelevinsky [135].

2

Theory of Symmetric Functions

In this section we discuss the algebra of symmetric functions, several of its bases, and interrelationships between them. The bases described below all have traditional names; those are monomial, elementary, complete homogeneous, and Schur symmetric functions, and also products of Newton power sums.

Take N variables $x_1, \dots, x_N$ and consider the algebra of polynomials $\mathbb{R}[x_1, \dots, x_N]$ over the field of real numbers. The polynomials which are invariant under permutations of the variables x_i generate a subalgebra

$$\mathrm{Sym}_N = \mathbb{R}[x_1, \dots, x_N]^{S(N)}.$$

This algebra is graded:

$$\mathrm{Sym}_N = \bigoplus_{n=0}^{\infty} \mathrm{Sym}_N^n,$$

where Sym_N^n consists of homogeneous symmetric polynomials of degree n.

For every multi-index $\alpha = (\alpha_1, \dots, \alpha_N) \in \mathbb{Z}_{\geq 0}^N$ denote by x^α the monomial $x_1^{\alpha_1} \cdots x_N^{\alpha_N}$. Let λ be an arbitrary partition with $\ell(\lambda) \leq N$. Clearly, the polynomials

$$m_\lambda = \sum_\alpha x^\alpha,$$

where the sum is taken over all distinct permutations of the entries of the vector $(\lambda_1, \lambda_2, \dots, \lambda_N)$, are symmetric, and they form a basis in Sym_N^n when λ ranges over all partitions of n of length $\leq N$.

For example,

$$m_{(1)} = x_1 + x_2 + \cdots, \qquad m_{(2)} = x_1^2 + x_2^2 + \cdots, \qquad m_{(1,1)} = \sum_{i<j} x_i x_j.$$

What we would like to do is to consider the symmetric polynomials in a very large number (or, even better, in an infinite number) of variables. The corresponding objects are called *symmetric functions*, and they can be formally constructed as follows.

For $N' > N$, define a projection (an algebra homomorphism)

$$\mathrm{pr}_{N',N} : \mathbb{R}[x_1, \dots, x_{N'}] \to \mathbb{R}[x_1, \dots, x_N]$$

which maps $x_{N+1}, \dots, x_{N'}$ to zero, and maps all the other variables $x_1, \dots, x_N$ just to themselves. Observe that

$$\mathrm{pr}_{N',N}(m_\lambda) = \begin{cases} m_\lambda, & \ell(\lambda) \le N, \\ 0, & \text{otherwise,} \end{cases}$$

where the "m_λ" in the left-hand size are polynomials in N' variables while those in the right-hand side are polynomials in N variables.

Let us now form the inverse (= projective) limit

$$\mathrm{Sym}^n = \varprojlim \mathrm{Sym}^n_N$$

with respect to projections $\mathrm{pr}_{N',N}$. This means that Sym^n is the space of sequences $(f_1, f_2, \dots)$ such that

$$f_N \in \mathrm{Sym}^n_N, \quad N = 1, 2, \dots; \qquad \mathrm{pr}_{N',N}(f_{N'}) = f_N, \quad N' > N.$$

Because of this stability property, every $f \in \mathrm{Sym}^n$ may be viewed as a true function on the set $\mathbb{R}^\infty_0$ formed by infinite real vectors $x = (x_1, x_2, \dots)$ with finitely many nonzero coordinates.

Example 2.1 Fix a partition λ. Then the sequence of symmetric polynomials m_λ in a growing number of variables is an element of Sym^n for $n = |\lambda|$; let us denote it still by m_λ. This element is called the *monomial symmetric function* corresponding to λ. It is not hard to show that $\{m_\lambda\}_{\lambda:\,|\lambda|=n}$ is a linear basis in Sym^n.

The *algebra of symmetric functions* over the base field $\mathbb{R}$ is, by definition, the direct sum

$$\mathrm{Sym} = \bigoplus_{n=0}^{\infty} \mathrm{Sym}^n, \qquad \mathrm{Sym}^0 := \mathbb{R}.$$

This is a graded unital algebra.

Example 2.2 The expression $\prod_{i=1}^{\infty}(1+x_i)$ does not define a symmetric function even though the symmetric polynomials $\prod_{i=1}^{N}(1+x_i)$ are consistent with projections $\mathrm{pr}_{N',N}$. Indeed, the degree of $\prod_{i=1}^{N}(1+x_i)$ grows with N while a symmetric function must be a *finite* linear combination of homogeneous components.

An equivalent definition of Sym is as follows: Elements of Sym are formal power series $f(x_1, x_2, \dots)$ in infinitely many indeterminates $x_1, x_2, \dots$, of bounded degree, and invariant under permutations of the x_is. (It does not matter whether the permutations in question are arbitrary or are assumed to move only finitely many indeterminates; this does not affect the definition.)

This definition makes it possible to work with infinitely many indeterminates, which is often very convenient. A general principle is that one deals with finitely or infinitely many x_is depending on the concrete situation. To distinguish between these two options, one can speak about symmetric *polynomials* or symmetric *functions*, respectively. The use of the term "functions" can be justified, e.g., by the fact that elements of Sym can be viewed as functions on $\mathbb{R}_0^\infty$.

The algebra of symmetric functions has a number of distinguished bases. One of them we have already introduced – it is the set of monomial symmetric functions parameterized by the partitions. Let us define another one.

For a partition of the form (1^r) (r nonzero parts all equal to 1), the corresponding *elementary symmetric function* is defined by

$$e_r = \sum_{i_1<i_2\cdots<i_r} x_{i_1}x_{i_2}\cdots x_{i_r} = m_{(1^r)}.$$

We also agree that $e_0 = 1$.

Alternatively, e_rs can be defined via their generating function

$$E(t) = \sum_{r=0}^{\infty} e_r t^r = \prod_{i=1}^{\infty}(1 + x_i t).$$

Note that if the number N of nonzero variables x_i is finite (thus, we are in Sym_N), then e_r vanishes for $r > N$.

For any partition $\lambda = (\lambda_1, \lambda_2, \ldots)$ we now define $e_\lambda = e_{\lambda_1}e_{\lambda_2}\cdots$. The fact that $\{e_\lambda\}$ is a basis of Sym is not as obvious as in the case of the monomial functions. To prove this fact we need some preparation.

Definition 2.3 The *lexicographic order* on partitions of the same number n is a linear order on $\mathbb{Y}_n$ defined as follows: $\mu \le \lambda$ if and only if $\mu = \lambda$ or else for some i

$$\mu_1 = \lambda_1, \ \ldots, \mu_i = \lambda_i, \ \mu_{i+1} < \lambda_{i+1}.$$

Example 2.4 The ordered set $\mathbb{Y}_5$ looks as follows:

$$(1, 1, 1, 1, 1) < (2, 1, 1, 1) < (2, 2, 1) < (3, 1, 1) < (3, 2) < (4, 1) < (5).$$

Proposition 2.5 *Let λ be a partition and λ' be its conjugate, that is, the corresponding Young diagrams are transposed with each other. Then*

$$e_{\lambda'} = m_\lambda + \sum_{\mu<\lambda} a_{\lambda\mu} m_\mu,$$

where $a_{\lambda\mu}$ are some nonnegative integers.

Example 2.6 For $\mathbb{Y}_1, \mathbb{Y}_2, \mathbb{Y}_3$ we have

$$\begin{aligned} e_1 &= m_{(1)} & e_{(1,1,1)} &= m_{(3)} + 3m_{(2,1)} + 6m_{(1,1,1)} \\ e_{(1,1)} &= m_{(2)} + 2m_{(1,1)} & e_{(2,1)} &= \qquad\quad m_{(2,1)} + 3m_{(1,1,1)} \\ e_2 &= \qquad\quad m_{(1,1)} & e_3 &= \qquad\qquad\qquad m_{(1,1,1)}. \end{aligned}$$

The triangular structure is clearly visible.

Proof of Proposition 2.5 The coefficient of m_λ in the expansion of any symmetric function in the basis of monomial functions equals the coefficient of x^λ in the expansion on monomials. Setting $m_i = \lambda_i - \lambda_{i+1}$ we have

$$e_{\lambda'} = e_1^{m_1} e_2^{m_2} \cdots = (x_1 + \cdots)^{m_1} (x_1 x_2 + \cdots)^{m_2} (x_1 x_2 x_3 + \cdots)^{m_3} \cdots,$$

where the dots in the parentheses denote lower terms in the lexicographic order. When we open the parentheses and list the resulting monomials in the lexicographic order, the highest one will be $x_1^{m_1}(x_1x_2)^{m_2}(x_1x_2x_2)^{m_3}\cdots = x^\lambda$. Hence, m_λ enters the expansion of $e_{\lambda'}$ in the basis of monomial symmetric functions with coefficient 1, and all other participating m_μs are smaller in the sense that $\mu < \lambda$. The fact that $a_{\lambda\mu} \in \mathbb{Z}_{\geq 0}$ is obvious. □

Proposition 2.5 immediately implies the following corollary.

Corollary 2.7 *The functions $\{e_r\}_{r\geq 1}$ are algebraically independent, and*

$$\mathrm{Sym} = \mathbb{R}[e_1, e_2, \ldots].$$

Likewise, $\mathrm{Sym}_N = \mathbb{R}[e_1, \ldots, e_N]$ (see Exercise 2.1, below).

Let us now define the third basis, $\{h_\lambda\}$, which is built from the *complete homogeneous symmetric functions* h_r. We set

$$h_0 = 1, \qquad h_r = h_{(r)} = \sum_{|\lambda|=r} m_\lambda = \sum_{i_1 \leq \cdots \leq i_r} x_{i_1} \cdots x_{i_r}, \qquad r \geq 1,$$

$$h_\lambda = h_{\lambda_1} h_{\lambda_2} \cdots \quad \text{for a partition} \quad \lambda = (\lambda_1, \lambda_2, \ldots).$$

Alternatively, the generating function for $\{h_r\}_{r\geq 0}$ has the form

$$H(t) = \sum_{r=0}^{\infty} h_r t^r = \prod_{i\geq 1} \frac{1}{1 - x_i t}.$$

Observe that the generating function $E(t)$ is just the inverse of $H(-t)$. That is, $H(-t)E(t) = 1$, which is equivalent to

$$\sum_{r=0}^{n} (-1)^r e_r h_{n-r} = 0, \qquad n = 1, 2, \ldots.$$

Since e_rs are algebraically independent, we may define an algebra homomorphism $\omega : \mathrm{Sym} \to \mathrm{Sym}$ by $\omega(e_r) = h_r$, $r \geq 1$. Since the above relations uniquely determine h_rs as polynomials in e_rs, and they do not change if we swap es and hs, we conclude that $\omega^2 = \mathrm{Id}$. Therefore, ω is an automorphism of the algebra Sym, and we obtain the following statement.

Proposition 2.8 *The complete homogeneous symmetric functions $h_1, h_2, \ldots$ are algebraically independent, and*

$$\mathrm{Sym} = \mathbb{R}[h_1, h_2, \ldots].$$

The next basis of Sym that we introduce is obtained from the *Newton power sums*

$$p_r = \sum_{i \geq 1} x_i^r = m_{(r)}, \quad r \geq 1.$$

The basis elements are again monomials labeled by arbitrary partitions $\lambda = (\lambda_1, \lambda_2, \ldots)$:

$$p_\lambda = \begin{cases} p_{\lambda_1} \cdots p_{\ell(\lambda)}, & \lambda \neq \varnothing, \\ 1, & \lambda = \varnothing. \end{cases}$$

A generating function for the power sums is defined by

$$P(t) = \sum_{r=1}^{\infty} \frac{p_r t^r}{r}.$$

In contrast to the generating series $H(t)$ and $E(t)$, the series $P(t)$ starts with a degree 1 term. This is one of the reasons why we avoid extending the definition of the Newton power sums p_r to $r = 0$; it seems that this would be incorrect, contrary to the case of the h_r and e_r where it is reasonable to set $h_0 = e_0 = 1$.

Proposition 2.9 *The functions $\{p_r\}_{r \geq 1}$ are algebraically independent, and*

$$\mathrm{Sym} = \mathbb{R}[p_1, p_2, \ldots].$$

Proof Observe that

$$P(t) = \sum_{i \geq 1} \sum_{r=1}^{\infty} \frac{x_i^r t^r}{r} = -\sum_{i \geq 1} \log(1 - x_i t) = -\log \prod_{i \geq 1} (1 - x_i t) = \log H(t).$$

Therefore,

$$P'(t) = \sum_{r=1}^{\infty} p_r t^{r-1} = \frac{H'(t)}{H(t)}.$$

Similarly, $P'(-t) = E'(t)/E(t)$. These relations imply

$$nh_n = \sum_{r=1}^{n} p_r h_{n-r}, \qquad ne_n = \sum_{r=1}^{n} (-1)^{r-1} p_r e_{n-r}, \qquad n \geq 1.$$

Hence, es and hs can be polynomially expressed through ps, and *vice versa*. □

For us the most important basis in Sym is formed by the *Schur functions*, which are defined as follows. Let us first assume that the number of variables is finite and denote it by N, as above. Take a monomial $x^\alpha = x_1^{\alpha_1} \cdots x_N^{\alpha_N}$ and antisymmetrize it:

$$a_\alpha = a_\alpha(x_1, \dots, x_N) := \sum_{\sigma \in S(N)} \operatorname{sgn} \sigma \cdot \sigma(x^\alpha), \tag{2.1}$$

where the action of a permutation σ on x^α consists in permuting the variables.

The polynomial a_α is skew-symmetric. In particular, if it is nonzero then the α_i are pairwise distinct. Thus, by possibly changing the sign of a_α, we may assume that $\alpha_1 > \alpha_2 > \cdots > \alpha_N \geq 0$, or, in other words, $\alpha = \lambda + \delta$, where λ is a partition with $\ell(\lambda) \leq N$ and

$$\delta = \delta_N = (N-1, N-2, \dots, 1, 0).$$

We have

$$a_{\lambda+\delta} = \det \left[x_i^{\lambda_j + N - j} \right]_{i,j=1}^{N}.$$

This determinant is divisible in the algebra of polynomials by each of the differences $x_i - x_j$ (because of the skew-symmetry). Hence, it is divisible by their product, which is the Vandermonde determinant. Set

$$s_\lambda(x_1, \dots, x_N) = \frac{\det \left[x_i^{\lambda_j + N - j} \right]_{i,j=1}^{N}}{\det \left[x_i^{N-j} \right]_{i,j=1}^{N}} = \frac{a_{\lambda+\delta}}{a_\delta}.$$

This is a symmetric polynomial called the *Schur polynomial* labeled by λ. We agree that if the number of variables is smaller then $\ell(\lambda)$ then $s_\lambda(x_1, \dots, x_N) \equiv 0$.

Proposition 2.10 *The polynomials* $\{s_\lambda\}_{\lambda:\, \ell(\lambda) \leq N}$ *form a basis in* Sym_N.

Proof Multiplication by a_δ is a linear isomorphism of Sym_N onto the space of skew-symmetric polynomials in $x_1, \dots, x_N$. On the other hand, $\{a_{\lambda+\delta}\}_{\ell(\lambda) \leq N}$ is a linear basis in this space. □

One readily checks that any Schur polynomial in $N+1$ variables turns into the corresponding polynomial in N variables by setting one of the variable to zero:

$$s_\lambda(x_1,\ldots,x_N,0) = s_\lambda(x_1,\ldots,x_N)$$

for any partition λ. This makes it possible to give the following definition.

Definition 2.11 For any partition (i.e., Young diagram) λ, the element of Sym defined by the sequence of Schur polynomials $\{s_\lambda(x_1,\ldots,x_N)\}_{N=0}^{\infty}$ is called the *Schur symmetric function* corresponding to λ and denoted by s_λ or $s_\lambda(x_1,x_2,\ldots)$. Since all polynomials $s_\lambda(x_1,\ldots,x_N)$ are homogeneous of degree $|\lambda|$, we have $s_\lambda \in \mathrm{Sym}^{|\lambda|}$.

We proceed with proving a few elementary formulas involving the Schur polynomials and the Schur functions.

Proposition 2.12 (Cauchy's Identity) *For any* $N=1,2,\ldots$

$$\prod_{i,j=1}^{N} \frac{1}{1-x_i y_j} = \sum_{\lambda:\,\ell(\lambda)\le N} s_\lambda(x_1,\ldots,x_N)s_\lambda(y_1,\ldots,y_N).$$

Proof The proof is based on the following formula for the *Cauchy determinant*:

$$\det\left[\frac{1}{1-x_i y_j}\right]_{i,j=1}^{N} = \frac{\prod\limits_{1\le i<j\le N}(x_i-x_j)(y_i-y_j)}{\prod\limits_{i,j=1}^{n}(1-x_i y_j)}. \tag{2.2}$$

(A hint for proving this formula can be found in Exercise 2.5, below.)

Substitution of this formula turns Cauchy's identity into

$$\det\left[\frac{1}{1-x_i y_j}\right]_{i,j=1}^{N} = \sum_{\lambda:\,\ell(\lambda)\le N} a_{\lambda+\delta}(x_1,\ldots,x_N)a_{\lambda+\delta}(y_1,\ldots,y_N).$$

Using the geometric series $(1-q)^{-1} = \sum_{k\ge 0} q^k$ for all the entries of the determinant, we obtain

$$\det\left[\frac{1}{1-x_i y_j}\right]_{i,j=1}^{N} = \sum_{k_1>\cdots>k_N\ge 0}\ \sum_{\sigma,\tau\in S(n)} \mathrm{sgn}(\sigma\tau)\cdot(x_{\sigma(1)}y_{\tau(1)})^{k_1}\cdots (x_{\sigma(n)}y_{\tau(n)})^{k_N},$$

and the right-hand side is readily seen to coincide with the needed sum over λ of $a_{\lambda+\delta}(x)a_{\lambda+\delta}(y)$. □

Remark 2.13 It is often convenient to omit the restriction $i, j \le N$ and write the Cauchy identity in the form

$$\prod_{i,j=1}^{\infty} \frac{1}{1 - x_i y_j} = \sum_{\lambda \in \mathbb{Y}} s_\lambda(x_1, x_2, \ldots) s_\lambda(y_1, y_2, \ldots),$$

which makes sense, e.g., as an identity of functions on $\mathbb{R}_0^\infty \times \mathbb{R}_0^\infty$.

Proposition 2.14 (Jacobi–Trudi formula) *For any partition λ, the following formula expressing the Schur symmetric function s_λ as a polynomial in one-row complete homogeneous symmetric functions $\{h_r\}$ holds:*

$$s_\lambda = \det\left[h_{\lambda_i - i + j}\right]_{i,j=1}^{N}.$$

Here N is an arbitrary integer $\ge \ell(\lambda)$, $h_0 := 1$, and $h_{-n} := 0$ for $n > 0$.

Proof It suffices to prove the formula for Schur polynomials with N variables. Observe that if $f(x) = \sum_{m \ge 0} f_m x^m$ is an arbitrary formal power series, then

$$f(x_1) \cdots f(x_N) = \sum_{\lambda:\, \ell(\lambda) \le N} \det\left[f_{\lambda_i - i + j}\right]_{i,j=1}^{N} s_\lambda(x_1, \ldots, x_N),$$

where we agree that $f_{-m} = 0$ for $m > 0$. Indeed, to prove this formula one just collects the coefficients of $x_1^{k_1} \cdots x_N^{k_N}$ in $a_\delta(x_1, \ldots, x_N) \cdot f(x_1) \cdots f(x_N)$.

It remains to apply this formula to

$$f(x) := \sum_{r=0}^{\infty} h_r(y_1, \ldots, y_N) x^r = \prod_{i=1}^{r} \frac{1}{1 - x y_i}$$

and to use Cauchy's identity. □

Here is another classical determinantal formula, which expresses general Schur functions through Schur functions parameterized by *hook Young diagrams*, that is, Young diagrams with the length of the diagonal equal to 1. Such diagrams are written in Frobenius notation as $(p \mid q)$.

Proposition 2.15 (Giambelli formula) *For any Young diagram λ with Frobenius coordinates $(p_1, \ldots, p_d \mid q_1, \ldots, q_d)$ one has*

$$s_\lambda = \det[s_{(p_i \mid q_j)}]_{i,j=1}^{d}.$$

Proof The proof can be found in, for example, Macdonald [72, Ex. I.3.9]. □

Next, let us discuss the interaction of Schur functions with power sums.

Definition 2.16 A *rim hook* is an edgewise connected set of boxes on the border of a Young diagram, which does not contain any 2×2 square. (See Sagan [107, Definition 4.10.1].)

Proposition 2.17 *For any partition μ and $r \geq 1$ one has*

$$p_r \cdot s_\mu = \sum_\lambda (-1)^{\text{height}(\lambda-\mu)} s_\lambda,$$

where the sum is taken over all Young diagrams $\lambda \supset \mu$ such that the complement $\lambda - \mu$ of μ in λ is a rim hook, and its height is the total number of rows it occupies minus 1.

Proof We have

$$p_r(x_1, \dots, x_N) \cdot a_{\mu+\delta}(x_1, \dots, x_N) = \sum_{k=1}^{N} a_{\mu+\delta+r\bar{\varepsilon}_k},$$

where $\bar{\varepsilon}_k \in \mathbb{Z}^N$ has the kth coordinate equal to 1 and all other coordinates equal to 0. Adding r to the kth row of $\mu + \delta$ so that it becomes the lth row of $\lambda + \delta$ for another partition λ is equivalent to adding a rim hook occupying rows from l to k to μ and obtaining λ. The change of sign from $a_{\mu+\delta+r\bar{\varepsilon}_k}$ to $a_{\lambda+\delta}$ comes from reordering the coordinates, and it equals $(-1)^{k-l} = (-1)^{\text{height}(\lambda-\mu)}$. □

Corollary 2.18 (Murnaghan–Nakayama rule) *Let ρ and λ be two partitions with $|\rho| = |\lambda|$. The coefficient of s_λ in the expansion of p_ρ in the basis of the Schur functions is equal to $\sum_S (-1)^{\text{height}(S)}$, where the sum is taken over all sequences of partitions*

$$S = \{(\varnothing = \lambda^{(0)} \subset \lambda^{(1)} \cdots \subset \lambda^{(\ell(\rho))} = \lambda)\}$$

such that $\lambda^{(i)} - \lambda^{(i-1)}$ is a rim hook with ρ_i boxes, and

$$\text{height}(S) := \sum_{i=1}^{\ell(\rho)} \text{height}(\lambda^{(i)} - \lambda^{(i-1)}).$$

Proof Induction on $\ell(\rho)$ using Proposition 2.17. □

We conclude this section by stating some results which relate the symmetric functions to the irreducible characters of the symmetric groups that we discussed in Chapter 1.

Definition 2.19 Consider a map $\psi : S(n) \to \text{Sym}^n$ defined by $\psi(\sigma) = p_{\rho(\sigma)}$, where the partition $\rho = \rho(\sigma)$ is the cycle structure of σ. For any $f \in \mathbb{R}[S(n)]$ define

$$\operatorname{ch}(f) = \frac{1}{n!} \sum_{\sigma \in S(n)} f(\sigma)\psi(\sigma).$$

The map $\operatorname{ch} : \mathbb{R}[S(n)] \to \operatorname{Sym}^n$ is called the *characteristic map*.

Proposition 2.20 (See Section I.7 in Macdonald [72]) *Let $\chi^\lambda \in \mathbb{R}[S(n)]$ be the trace of the irreducible representation of $S(n)$ corresponding to $\lambda \in \mathbb{Y}_n$. Then*

$$\operatorname{ch}(\chi^\lambda) = s_\lambda.$$

This statement allows us to express the transition matrix between the basis of the Schur functions and the basis of the power sums, as well as its inverse, in terms of the irreducible characters of the symmetric groups. Indeed, the image of the indicator function of a conjugacy class C_ρ of $S(n)$ parameterized by $\rho \in \mathbb{Y}_n$ under the characteristic map is $|C_\rho| p_\rho / n!$. This immediately gives

$$s_\lambda = \frac{1}{n!} \sum_{\rho \in \mathbb{Y}_n} \chi_\rho^\lambda |C_\rho| \cdot p_\rho, \qquad \lambda \in \mathbb{Y}_n,$$

where χ_ρ^λ stands for the value of χ^λ on any of the elements in C_ρ. Using Macdonald's notation

$$z_\rho = \frac{n!}{|C_\rho|}$$

we rewrite the above relation as

$$s_\lambda = \sum_{\rho \in \mathbb{Y}_n} z_\rho^{-1} \chi_\rho^\lambda p_\rho, \qquad \lambda \in \mathbb{Y}_n. \tag{2.3}$$

An explicit expression for z_ρ is given in Exercise 2.14.

Applying the orthogonality relations for the characters of $S(n)$ (see, e.g., Sagan [107, Section 1.9]), we also get the following statement.

Proposition 2.21 (Frobenius' formula) *For any partition ρ*

$$p_\rho = \sum_{\lambda \in \mathbb{Y}_{|\rho|}} \chi_\rho^\lambda \cdot s_\lambda.$$

Frobenius' formula is the most efficient theoretical tool for handling the characters. One of its consequences is that the Murnaghan–Nakayama rule (Corollary 2.18) provides an algorithm for computing the character values χ_ρ^λ.

2.1 Exercises

Exercise 2.1 Deduce from the proof of Proposition 2.5 that the elements $e_1, \dots, e_N$, viewed as symmetric polynomials in N variables, are algebraically independent and generate the algebra Sym_N. This result is often referred to as the "fundamental theorem of symmetric polynomials".

Exercise 2.2 Denote by $\mathbb{R}_1^\infty$ the space of infinite sequences $(x_1, x_2, \dots)$ of real numbers such that $\sum_{i=1}^\infty |x_i| < \infty$. Show that the elements of Sym can be correctly defined as functions on $\mathbb{R}_1^\infty$.

Exercise 2.3 (a) Take $\lambda \in \mathbb{Y}_n$ and consider the expansion of e_λ in the basis of monomial functions:

$$e_\lambda = \sum_{\mu \in \mathbb{Y}_n} M_{\lambda\mu} m_\mu.$$

Prove that the coefficient $M_{\lambda\mu}$ is equal to the number of matrices $A = [a_{ij}]$ which are large enough (e.g., of size $n \times n$), and which satisfy for every i, j the constraints

$$a_{ij} \in \{0, 1\}, \qquad \sum_k a_{ik} = \lambda_i, \qquad \sum_k a_{kj} = \mu_j.$$

(b) Similarly, prove that the coefficient $N_{\lambda\mu}$ in the expansion

$$h_\lambda = \sum_{\mu \in \mathbb{Y}_n} N_{\lambda\mu} m_\mu$$

is equal to the number of large enough matrices $B = [b_{ij}]$ such that

$$b_{ij} \in \{0, 1, 2, \dots\}, \qquad \sum_k b_{ik} = \lambda_i, \qquad \sum_k b_{kj} = \mu_j.$$

(The only difference between A and B is in the range of their matrix elements.)

Exercise 2.4 Show that the image of the power sum p_r under involution ω is $(-1)^{r-1} p_r$.

Exercise 2.5 (a) Prove formula (2.2) for the Cauchy determinant. (Hint: use the fact that a polynomial in x_is and y_js, which is separately skew-symmetric with respect to each set of variables, is divisible by the product of the Vandermonde determinant in x_is and the Vandermonde determinant in y_js.)

(b) Prove the following generalization of the Cauchy determinant formula: For $M \le N$

$$\frac{\prod_{1\le i<j\le M}(x_i - x_j)\prod_{1\le i<j\le N}(y_i - y_j)}{\prod_{i=1}^M\prod_{j=1}^N(x_i + y_j)}$$
$$= \sum_{y'\sqcup y''=y} \operatorname{sgn}(y', y'') \prod_{1\le i<j\le N-M} (y'_i - y'_j) \cdot \det\left[\frac{1}{x_i + y''_j}\right]_{i,j=1}^M .$$

Here the summation in the right-hand side is taken over all possible ways to split the set $y = \{y_j\}_{j=1}^N$ into two disjoint sets $y' = \{y'_1, \dots, y'_{N-M}\}$ and $y'' = \{y''_1, \dots, y''_M\}$, and $\operatorname{sgn}(y', y'')$ stands for the sign of the permutation that brings the sequence $(y'_1, \dots, y'_{N-M}, y''_1, \dots, y''_M)$ to the sequence $(y_1, \dots, y_N)$.

Exercise 2.6 Prove the *dual Cauchy identity*

$$\prod_{i,j}(1 + x_i y_j) = \sum_\lambda s_\lambda(x_1, x_2, \dots)s_{\lambda'}(y_1, y_2, \dots).$$

Exercise 2.7 Prove the *dual Jacobi–Trudi formula*: For any partition λ and $N \ge \ell(\lambda')$

$$s_\lambda = \det[e_{\lambda'_i - i + j}]_{i,j=1}^N,$$

where $e_0 := 1$ and $e_{-m} := 0$ for $m > 0$.

Exercise 2.8 Deduce from the above formula the *duality relation* for the Schur functions:

$$\omega(s_\lambda) = s_{\lambda'}.$$

Exercise 2.9 Prove *Pieri's formula*: For a partition μ and $r \ge 1$

$$h_r s_\mu = \sum_\lambda s_\lambda,$$

where the sum is taken over all partitions (= Young diagrams) λ such that $\lambda - \mu$ is a *horizontal r-strip* (meaning that $\lambda - \mu$ has r boxes no two of which lie in the same column). Dually,

$$e_r s_\mu = \sum_\lambda s_\lambda,$$

where the sum is taken over all λ such that $\lambda - \mu$ is a *vertical r-strip* (no two boxes are in the same row).

Exercise 2.10 Prove the *binomial formula*

$$s_\lambda(1 + x_1, \dots, 1 + x_N) = \sum_{\mu:\, \mu\subset\lambda} \det\left[\binom{\lambda_i + N - i}{\mu_j + N - j}\right]_{i,j=1}^N s_\mu(x_1, \dots, x_N).$$

Here the sum is taken over all Young diagrams μ which are subsets of the Young diagram λ.

Exercise 2.11 Show that for any $n \geq 1$,

$$p_1^n = \sum_{\lambda \in \mathbb{Y}_n} \dim \lambda \cdot s_\lambda.$$

Exercise 2.12 Apply the Murnaghan–Nakayama rule to:

(a) evaluate $\chi^{(p|q)}_{(p+q+1)}$;

(b) prove that $\dim \lambda$ equals the number of standard Young tableaux of shape λ;

(c) prove that $\chi^\lambda_\rho = 0$ if $d(\lambda) > \ell(\rho)$.

Exercise 2.13 (a) Let $\chi_1 \in \mathbb{R}[S(n_1)]$ and $\chi_2 \in \mathbb{R}[S(n_2)]$ be the traces of (finite-dimensional) representations π_1 and π_2 of $S(n_1)$ and $S(n_2)$, respectively. Prove that

$$\mathrm{ch}\left(\text{trace of } \mathrm{Ind}^{S(n_1+n_2)}_{S(n_1)\times S(n_2)}(\pi_1 \otimes \pi_2)\right) = \mathrm{ch}(\chi_1)\,\mathrm{ch}(\chi_2).$$

(Hint: use Frobenius reciprocity.)

(b) Let $c^\lambda_{\mu\nu}$ denote the coefficient of s_λ in the expansion of $s_\mu s_\nu$ in the basis of the Schur functions. Clearly, $c^\lambda_{\mu\nu}$ can be nonzero only if $|\lambda| = |\mu| + |\nu|$. The coefficients $c^\lambda_{\mu\nu}$ are called the *Littlewood–Richardson coefficients*. Apply (a) to show that $c^\lambda_{\mu\nu}$ is equal to the multiplicity of the irreducible representation π_λ in $\mathrm{Ind}^{S(|\lambda|)}_{S(|\mu|)\times S(|\nu|)}(\pi_\mu \otimes \pi_\nu)$. Thus, $c^\lambda_{\mu\nu}$ is a nonnegative integer. (This fact is used below in the proof of Theorem 4.3.)

Exercise 2.14 Recall the notation

$$z_\rho = \frac{n!}{|C_\rho|},$$

where ρ is a partition, $n = |\rho|$, and C_ρ is the conjugacy class in $S(n)$ corresponding to ρ. Show that

$$z_\rho = \prod_{i \geq 1} i^{m_i(\rho)} m_i(\rho)!,$$

where $m_i(\rho)$ stands for the number of parts in ρ equal to i.

Exercise 2.15 Equip Sym with the inner product $(\cdot\,,\,\cdot)$ in which the Schur functions form an orthonormal basis.

(a) Show that the functions p_ρ, where ρ ranges over the set of partitions, form an orthogonal basis and

$$(p_\rho, p_\rho) = z_\rho,$$

where the quantities z_ρ are defined in Exercise 2.14.

(b) Let $\{a_\lambda\}$ be an arbitrary homogeneous basis in Sym (here λ may range over an abstract set of indices but in concrete applications this will always be the set of partitions). Observe that there exists a unique biorthogonal homogeneous basis $\{b_\lambda\}$, that is, $(a_\lambda, b_\mu) = \delta_{\lambda\mu}$. Show that

$$\prod_{i,j=1}^{\infty} \frac{1}{1 - x_i y_j} = \sum_{\lambda \in \mathbb{Y}} a_\lambda(x_1, x_2, \ldots) b_\lambda(y_1, y_2, \ldots).$$

Conversely, if $\{a_\lambda\}$ and $\{b_\lambda\}$ are two homogeneous bases in Sym such that the above identity holds then these are biorthogonal bases.

(c) Show that $\{m_\lambda\}$ and $\{h_\lambda\}$ are biorthogonal bases.

Exercise 2.16 Define the algebra morphism Δ : Sym $\to$ Sym $\otimes$ Sym by setting

$$\Delta(1) = 1, \qquad \Delta(p_k) = p_k \otimes 1 + 1 \otimes p_k, \quad k = 1, 2, \ldots.$$

The map Δ is called the *comultiplication* in Sym. (Together with a natural counit map Sym $\to \mathbb{R}$, Δ satisfies a number of conditions meaning that Sym is a Hopf algebra, but we will not use this fact.)

(a) Let us fix an arbitrary splitting of the set $x_1, x_2, \ldots$ of the indeterminates into two disjoint infinite parts, say $x_1', x_2', \ldots$ and $x_1'', x_2'', \ldots$. Let $f \in$ Sym be an arbitrary element and write $\Delta(f)$ as a finite sum

$$\Delta(f) = \sum_i g_i' \otimes g_i'', \qquad g_i', g_i'' \in \text{Sym},$$

which is always possible. Show that

$$f(x_1, x_2, \ldots) = \sum_i g_i'(x_1', x_2', \ldots) g_i''(x_1'', x_2'', \ldots).$$

Thus, the comultiplication map has a very simple interpretation: we regard symmetric functions as separately symmetric ones, with respect to a splitting of variables into two groups.

(b) Show that

$$\Delta(h_k) = h_1 \otimes h_{k-1} + h_2 \otimes h_{k-2} + \cdots + h_{k-1} \otimes h_1.$$

In terms of the generating series $H(t)$ this can be conveniently written as

$$\Delta(H(t)) = H(t) \otimes H(t).$$

(c) Show that the action of Δ on the elementary symmetric functions is given by the similar formula,

$$\Delta(E(t)) = E(t) \otimes E(t).$$

(d) Let $\{a_\lambda\}$ and $\{b_\lambda\}$ be arbitrary biorthogonal homogeneous bases in Sym. Show that the structure constants of the comultiplication map in the the basis $\{a_\lambda\}$ coincide with the structure constants of the multiplication map in the basis $\{b_\lambda\}$. As a corollary, one gets the following important fact: the structure constants of Δ in the basis of Schur functions coincide with the Littlewood–Richardson coefficients $c^\lambda_{\mu\nu}$ (see Exercise 2.13):

$$\Delta(s_\lambda) = \sum_{\substack{\mu,\nu \\ |\mu|+|\nu|=|\lambda|}} c^\lambda_{\mu\nu} s_\mu \otimes s_\nu.$$

In particular, they are nonnegative integers.

(e) Show that the structure constants of Δ in the basis of monomial symmetric functions are nonnegative integers, too.

Exercise 2.17 The *algebra of supersymmetric functions* is defined as the subalgebra in Sym ⊗ Sym that is the image of Sym under the morphism

$$(\mathrm{id} \otimes \omega) \circ \Delta : \mathrm{Sym} \to \mathrm{Sym} \otimes \mathrm{Sym},$$

where $\omega : \mathrm{Sym} \to \mathrm{Sym}$ is the involution introduced just before Proposition 2.8 and Δ is the comultiplication map defined above. Since Δ is injective (prove this!), the above morphism is also injective, so that the algebra of supersymmetric functions is nothing other than a certain realization of the algebra Sym. However, this "super" realization is very useful.

(a) Viewing Sym ⊗ Sym as an algebra of functions in a doubly infinite collection of variables, say, $x = (x_i)$ and $y = (y_j)$, one can also describe the algebra of supersymmetric functions as the algebra generated by the *super power sums*

$$\begin{aligned} p_k(x; y) :&= \sum_i x_i^k + (-1)^{k-1} \sum_j y_j^k \\ &= \sum_i x_i^k - \sum_j (-y_j)^k \end{aligned}$$

(indeed, the equivalence of the both definitions follows from Exercise 2.4).[1]

(b) For an arbitrary element $f \in \mathrm{Sym}$ we will denote by $f(x; y)$ the corresponding supersymmetric function. Show that

$$(\omega(f))(x; y) = f(y; x).$$

As a consequence, one obtains (see Exercise 2.8),

$$s_{\lambda'}(x; y) = s_\lambda(y, x).$$

[1] Some authors (including Macdonald [72]) adopt a slightly different definition of supersymmetric functions, which reduces to our definition by changing the sign of y_js.

(c) Show that the "super" version of the generating series $H(t)$ and $E(t)$ for the complete and elementary symmetric functions takes the form

$$H(t)(x;y) = \frac{\prod(1+y_j t)}{\prod(1-x_i t)}, \qquad E(t)(x;y) = \frac{\prod(1+x_i t)}{\prod(1-y_j t)}.$$

2.2 Notes

There are several excellent treatments of symmetric functions available in the literature, see, e.g., Macdonald [72], Stanley [113], Sagan [107]. We mostly followed Macdonald [72] in our brief exposition.

3

Coherent Systems on the Young Graph

3.1 The Infinite Symmetric Group and the Young Graph

Definition 3.1 The *infinite symmetric group* $S(\infty)$ is the group of finite permutations of the set $\mathbb{Z}_{>0} = \{1, 2, \ldots\}$. In other words, an element of $S(\infty)$ is a bijection $\sigma : \mathbb{Z}_{>0} \to \mathbb{Z}_{>0}$ such that $\sigma(x) = x$ for x large enough.

We fix a tower of finite symmetric groups

$$S(1) \subset S(2) \subset \cdots \subset S(n) \subset \cdots$$

so that the union of them is $S(\infty)$. The simplest way to do it is to realize $S(n)$ as permutations of the subset $\{1, \ldots, n\} \subset \mathbb{Z}_{>0}$. That is, we embed $S(n)$ into $S(\infty)$ as the subgroup of bijections that are identical outside $\{1, \ldots, n\}$.

One can also say that $S(\infty)$ is the direct (inductive) limit $\varinjlim S(n)$ of the finite symmetric groups with respect to embeddings $S(n) \to S(n+1)$ described above.

Definition 3.2 Let us denote by $\mathbb{Y}$ the set of all partitions: $\mathbb{Y} = \sqcup_{n \geq 0} \mathbb{Y}_n$. It is convenient to turn this set into a $\mathbb{Z}_{\geq 0}$-graded graph whose nth level consists of $\mathbb{Y}_n$, and whose edges join Young diagrams if they differ by exactly one box. It is called the *Young graph* and it is denoted by the same symbol $\mathbb{Y}$. See the picture below.

A general definition of what we mean by a graded graph is given in Chapter 7 (Definition 7.4).

The graph structure on the set $\mathbb{Y}$ reflects the Young branching rule (Proposition 1.4): two vertices $\mu \in \mathbb{Y}_{n-1}$ and $\lambda \in \mathbb{Y}_n$ are joined by an edge if and only if π_μ enters the decomposition of π_λ restricted to $S(n-1) \subset S(n)$.

The Young graph comes with a canonical collection of stochastic matrices that are indexed by numbers $n = 1, 2, \ldots$ and determine "transitions" from the nth level of the graph to the $(n-1)$th one. Here is the definition:

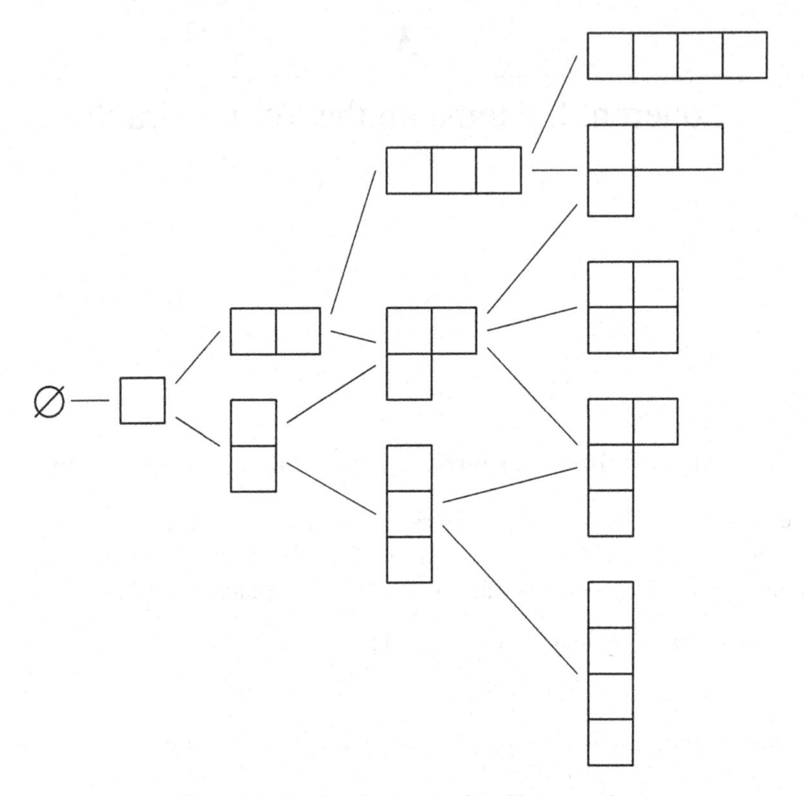

Figure 3.1 The first five levels of the Young graph.

Definition 3.3 Given $n = 1, 2, \ldots$, let λ range over $\mathbb{Y}_n$ and μ range over $\mathbb{Y}_{n-1}$. We define a $\mathbb{Y}_n \times \mathbb{Y}_{n-1}$ stochastic matrix $\Lambda^n_{n-1} = [\Lambda^n_{n-1}(\lambda, \mu)]$ by setting

$$\Lambda^n_{n-1}(\lambda, \mu) = \begin{cases} \dfrac{\dim \mu}{\dim \lambda}, & \mu \nearrow \lambda, \\ 0, & \text{otherwise.} \end{cases} \tag{3.1}$$

Recall that $\dim \lambda$ denotes the dimension of π_λ, or, in combinatorial terms, the number of standard Young tableaux of shape λ (see Exercise 2.12).

Recall also that a matrix is said to be *stochastic* if its entries are nonnegative and all row sums equal 1. For Λ^n_{n-1}, the first condition is obvious, and the second condition is equivalent to the equality

$$\sum_{\mu:\, \mu \nearrow \lambda} \dim \mu = \dim \lambda, \qquad \lambda \in \mathbb{Y}_n,$$

which follows from the Young branching rule.

3.2 Coherent Systems

We start the study of characters of $S(\infty)$ in the sense of Definition 1.7. Our first aim is to give for them a purely combinatorial interpretation in terms of the Young graph.

Observe that if M is a probability measure on $\mathbb{Y}_n$, viewed as a row vector with coordinates indexed by elements of $\mathbb{Y}_n$, then the row vector $M\Lambda^n_{n-1}$ is a probability measure on $\mathbb{Y}_{n-1}$. Indeed, this is an immediate consequence of the fact that Λ^n_{n-1} is a stochastic matrix.

Definition 3.4 We say that a sequence $\{M^{(n)}\}_{n\geq 0}$ of probability measures on the sets $\mathbb{Y}_n$ forms a *coherent system* of distributions on the Young graph if

$$M^{(n)}\Lambda^n_{n-1} = M^{(n-1)}, \qquad n = 1, 2, \dots . \tag{3.2}$$

In more detail, the above coherence relation means that

$$\sum_{\lambda:\lambda\searrow\mu} M^{(n)}(\lambda)\Lambda^n_{n-1}(\lambda,\mu) = M^{(n-1)}(\mu), \qquad n = 1, 2, \dots, \qquad \mu \in \mathbb{Y}_{n-1}. \tag{3.3}$$

Given a character χ of $S(\infty)$, we write χ_n for its restriction to the subgroup $S(n) \subset S(\infty)$.

Proposition 3.5 *There is a natural bijective correspondence $\chi \leftrightarrow \{M^{(n)}\}_{n\geq 0}$ between the characters of $S(\infty)$ and the coherent systems of probability distributions on $\mathbb{Y}$, uniquely determined by the relations*

$$\chi_n = \sum_{\lambda\in\mathbb{Y}_n} M^{(n)}(\lambda) \cdot \frac{\chi^\lambda}{\dim\lambda}, \qquad n = 1, 2, \dots . \tag{3.4}$$

Proof Again, this is a simple consequence of the Young branching rule.

Indeed, let χ be a character of $S(\infty)$. For every $n = 1, 2, \dots, \chi_n$ is a character of $S(n)$, as it follows from the very definition of characters. By Proposition 1.6, χ_n is uniquely written as a convex combination of the normalized irreducible characters $\chi^\lambda/\dim\lambda$, where λ ranges over $\mathbb{Y}_n$. The coefficients of this expansion form a probability distribution on $\mathbb{Y}_n$; denote it by $M^{(n)}$. These distributions obey the coherence relation (3.3), for it just expresses the fact that χ_{n-1} coincides with the restriction of χ_n to the subgroup $S(n-1) \subset S(n)$; here we use the Young branching rule.

Conversely, the above argument can be inverted: starting with a coherent system $\{M^{(n)}\}$ we define the characters χ_n of the groups $S(n)$ by means of (3.4). Then the coherence property says that χ_n is an extension of χ_{n-1}, so that the sequence $\{\chi_n\}$ comes from a character χ of $S(\infty)$. □

An obvious but important observation is that both the set of characters and the set of coherent systems have a natural structure of convex sets, and the bijection of Proposition 3.5 preserves this structure.

As was pointed out in Proposition 1.6, the set of characters of a finite group is isomorphic, as a convex set, to a finite-dimensional simplex. It turns out that a similar assertion holds for characters of the group $S(\infty)$. Namely, we will see that the convex set of characters of $S(\infty)$ can be realized as a "generalized simplex", that is, the set of probability measures on a certain space. In the next section we explain what this space is.

Definition 3.6 A function ψ on the vertices of the Young graph $\mathbb{Y}$ is said to be *harmonic* (in the sense of Vershik and Kerov; see their works [122], [126], [63]) if the value of ψ at each vertex is equal to the sum of the values at all the adjacent vertices located on the next level:

$$\psi(\mu) = \sum_{\lambda:\lambda\searrow\mu} \psi(\lambda). \tag{3.5}$$

This relation is essentially equivalent to the coherence relation (3.3). Namely, (3.3) just means that the function

$$\psi(\lambda) := \frac{M^{(n)}(\lambda)}{\dim \lambda}, \qquad n = |\lambda|, \tag{3.6}$$

is harmonic. It is easily seen that (3.6) determines a bijective correspondence between coherent systems and nonnegative harmonic functions taking value 1 at the root vertex $\varnothing \in \mathbb{Y}$. Due to the simplicity of relation (3.5) it is sometimes convenient to switch from the language of coherent systems to that of harmonic functions.

Remark 3.7 We would like to notice that the use of the term "harmonic function" in the present context disagrees with its conventional use in probabilistic potential theory and random walks on discrete groups: as we will see in Chapter 7, coherent systems of distributions (and hence harmonic functions in the Vershik–Kerov sense) are tied to the *entrance* boundaries of some Markov processes, while harmonic functions in the conventional sense are tied to the *exit* boundaries. On a discrete space, one often identifies functions and measures; Vershik–Kerov's harmonic functions should actually be viewed as (densities of) *measures*.

3.3 The Thoma Simplex

Let $[0, 1]^\infty$ be the space of infinite vectors $(x_1, x_2, \dots)$ with coordinates in the closed unit interval $[0, 1]$. We regard $[0, 1]^\infty$ as the direct product of countably many copies of $[0, 1]$ and equip it with the product topology, in which it is a compact metrizable separable space.

Definition 3.8 The *Thoma simplex* is the subset $\Omega \subset [0, 1]^\infty \times [0, 1]^\infty$ formed by pairs $\omega = (\alpha, \beta)$ of infinite vectors whose coordinates $\alpha_i \in [0, 1]$, $\beta_i \in [0, 1]$ satisfy the constraints

$$\alpha_1 \geq \alpha_2 \geq \cdots, \qquad \beta_1 \geq \beta_2 \geq \cdots, \qquad \sum_{i=1}^{\infty} (\alpha_i + \beta_i) \leq 1.$$

The above constraints just mean that the numbers $x_1, x_2, \dots, y_1, y_2, \dots,$ and γ defined by

$$x_i := i(\alpha_i - \alpha_{i+1}), \quad y_i := i(\beta_i - \beta_{i+1}), \quad \gamma := 1 - \sum_{i=1}^{\infty} (\alpha_i + \beta_i)$$

are nonnegative and sum to 1. Thus, Ω can be realized as a simplex with countably many vertices and coordinates $(x_1, x_2, \dots; y_1, y_2, \dots; \gamma)$, which explains the use of the word "simplex". However, the simplex structure on Ω is of no importance to us. What is really important is that Ω is closed in $[0, 1]^\infty \times [0, 1]^\infty$ and hence is itself a compact metrizable separable space.

Let $C(\Omega)$ denote the space of real-valued continuous functions on Ω with supremum norm; this is a real Banach algebra with respect to pointwise multiplication. Recall that $p_1, p_2, \dots$ denote the power-sum elements in Sym, which generate this algebra.

Proposition 3.9 (i) *There exists a unique algebra morphism $f \mapsto f^\circ$ of the algebra* Sym *into the algebra $C(\Omega)$, preserving the unit elements and such that*

$$p_k^\circ(\omega) = \sum_{i=1}^{\infty} \alpha_i^k + (-1)^{k-1} \sum_{i=1}^{\infty} \beta_i^k, \quad k = 2, 3, \dots, \tag{3.7}$$

where $\omega = (\alpha, \beta) \in \Omega$, and

$$p_1^\circ \equiv 1. \tag{3.8}$$

(ii) *The image of the algebra* Sym *in $C(\Omega)$ is dense in the norm topology.*

(iii) *The kernel of the morphism $f \mapsto f^\circ$ is the principal ideal in* Sym *generated by $p_1 - 1$.*

Proof (i) The uniqueness claim is evident because the elements $p_1, p_2, \dots$ are algebraically independent and generate the algebra Sym. To prove that the map $f \mapsto f^\circ$ is well defined we have to check that the functions p_k° defined by (3.7) and (3.8) are continuous. For $k = 1$ the assertion is trivial, so that we may assume $k \geq 2$.

Since α_is decrease, the condition $\sum \alpha_i \leq 1$ implies $\alpha_i \leq i^{-1}$ for any $i = 1, 2, \dots$, so that $\alpha_i^k \leq i^{-k}$. Similarly, $\beta_i^k \leq i^{-k}$. Since the series $\sum i^{-k} < \infty$ converges for $k \geq 2$, it follows that the two series in (3.7) converge uniformly on α and β, which implies the desired continuity property.

(ii) Observe that the nonzero values of α_is and β_is are uniquely reconstructed as the positive and negative poles of the generating series

$$\sum_{k=1}^{\infty} \frac{p_k^\circ(\alpha, \beta)}{u^k} = \sum_{i=1}^{\infty} \frac{\alpha_i}{u - \alpha_i} + \sum_{i=1}^{\infty} \frac{\beta_i}{u + \beta_i} + \frac{\gamma}{u},$$

where (we recall)

$$\gamma = 1 - \sum_{i=1}^{\infty} (\alpha_i + \beta_i). \tag{3.9}$$

Next, possible multiplicities of α_i and β_i are uniquely determined from the residues.

It follows that the functions $p_1^\circ \equiv 1$ and $p_k^\circ \in C(\Omega)$, $k \geq 2$, defined above separate points of Ω; then the desired claim will follow from the Stone–Weierstrass theorem whose formulation is given below.

(iii) By virtue of (3.8), the kernel of morphism $f \mapsto f^\circ$ contains the ideal generated by $p_1 - 1$. To prove that the kernel coincides with this ideal it remains to check that the functions $p_k^\circ \in C(\Omega)$ with $k \geq 2$ are algebraically independent. From (3.7) it is seen that these functions coincide with the conventional power sums in variables $\alpha_1, \alpha_2, \dots$ when restricted to the subset of Ω where all coordinates β_i vanish. Then the claim becomes evident. □

For the reader's convenience we formulate the Stone–Weierstrass theorem that we have used above. For its proof, see, e.g., Reed and Simon [105, Theorem IV.9].

Theorem 3.10 (Stone–Weierstrass theorem: real version) *Let X be a compact topological space, let $C(X)$ be the algebra of real-valued continuous functions on X with the supremum norm, and let $A \subset C(X)$ be a closed subalgebra in $C(X)$ that separates points. (That is, for any two distinct points x_1 and x_2 in X, there exists $f \in A$ such that $f(x_1) \neq f(x_2)$.) Furthermore, assume that A contains the constant function* 1. *Then $A = C(X)$.*

Remark 3.11 Let us explain why we have defined the functions p_k° separately for $k \geq 2$ and for $k = 1$, because, at first glance, (3.8) may look strange. A reason is that the expression $\sum \alpha_i + \sum \beta_i$ is not a continuous function. Here is a justification of (3.8): on the dense subset $\Omega_0 \subset \Omega$ (see Exercise 3.3), one can use (3.7) as the definition of the functions p_k° for all $k = 1, 2, \ldots$ and then extend the functions to the whole space Ω by continuity. Then for $k \geq 2$ we recover the initial definition, while for $k = 1$ we get the constant 1.

Formulas (3.7) and (3.8) extend by multiplicativity to arbitrary elements of the basis $\{p_\rho\}$ in Sym indexed by arbitrary partitions $\rho = (\rho_1, \ldots, \rho_\ell)$: we simply have

$$p_\rho^\circ(\omega) = p_{\rho_1}^\circ(\omega) \cdots p_{\rho_\ell}^\circ(\omega), \qquad \omega \in \Omega. \tag{3.10}$$

Then, by virtue of Frobenius' formula (more precisely, see its inversion (2.3)), we get

$$s_\lambda^\circ(\omega) = \sum_{\rho \vdash n} z_\rho^{-1} \chi_\rho^\lambda p_\rho^\circ(\omega), \qquad \lambda \in \mathbb{Y}, \quad n := |\lambda|, \quad \omega \in \Omega. \tag{3.11}$$

Note also that

$$s_\varnothing^\circ(\omega) \equiv s_{(1)}^\circ(\omega) \equiv 1. \tag{3.12}$$

Equivalently, the function $s_\lambda^\circ(\omega)$ can be written without use of the characters of $S(n)$ if, instead of the Frobenius formula, we apply the Jacobi–Trudi formula (Proposition 2.14):

$$s_\lambda^\circ(\omega) = \det\left[h_{\lambda_i - i + j}^\circ\right]_{i,j=1}^N, \qquad \omega \in \Omega, \tag{3.13}$$

where, as in Proposition 2.14, N is an arbitrary integer $\geq \ell(\lambda)$, $h_0(\omega) \equiv 1$, and $h_{-k}(\omega) \equiv 0$ for $k > 0$. In turn, the functions $h_n^\circ(\omega)$ are defined through a generating series in a formal variable t:

$$1 + \sum_{k=1}^\infty h_k^\circ(\omega) t^k = e^{\gamma t} \prod_{i=1}^\infty \frac{1 + \beta_i t}{1 - \alpha_i t}, \qquad \omega = (\alpha, \beta) \in \Omega. \tag{3.14}$$

We recall that

$$\gamma = 1 - \sum_{i=1}^\infty (\alpha_i + \beta_i).$$

The proof of (3.14) is easy and is left as an exercise; see Exercise 3.4.

3.4 Integral Representation of Coherent Systems and Characters

Here we state the main results of Part I. Recall that the group $S(\infty)$ is introduced in Definition 3.1, its characters are understood in accordance with general Definition 1.7, the notion of a coherent system is explained in Definition 3.4, the Thoma simplex is introduced in Definition 3.8, and the functions $s_\lambda^\circ(\omega)$ on Ω, indexed by arbitrary Young diagrams λ, can be defined by equivalent formulas (3.11) and (3.13), which come from the general definition of the map $\mathrm{Sym} \to C(\Omega)$ introduced in Proposition 3.9.

Theorem 3.12 (Integral representation of coherent systems) *There is a bijective correspondence $\{M^{(n)}\} \leftrightarrow P$ between the coherent systems of distributions on the Young graph $\mathbb{Y}$ and the probability Borel measures on the Thoma simplex Ω, determined by the formula*

$$M^{(n)}(\lambda) = \dim\lambda \int_\Omega s_\lambda^\circ(\omega)P(d\omega), \qquad n = 0, 1, 2, \dots, \qquad \lambda \in \mathbb{Y}_n. \quad (3.15)$$

Dropping the prefactor $\dim\lambda$ we get the integral representation for nonnegative normalized harmonic functions on $\mathbb{Y}$ (Definition 3.6):

$$\psi(\lambda) = \int_\Omega s_\lambda^\circ(\omega)P(d\omega), \quad \lambda \in \mathbb{Y}. \quad (3.16)$$

The proof of Theorem 3.12 is postponed to Chapter 6.

Let us introduce a definition which will be used later on.

Definition 3.13 We call P the *boundary measure* of the corresponding coherent system $\{M^{(n)}\}$.

Now, we aim to restate the assertion of the theorem in terms of characters. To do this we need a parameterization of conjugacy classes in $S(\infty)$, which is easily obtained by analogy with the case of finite symmetric groups $S(n)$:

Proposition 3.14 *The conjugacy classes in $S(\infty)$ can be parameterized by partitions ρ without parts equal to* 1.

Proof Using the fact that any finite number of elements of $S(\infty)$ is contained in the subgroup $S(n)$ for n large enough, one easily shows that two given elements of $S(\infty)$ are in the same conjugacy class if and only if the (unordered) collections of lengths of their nontrivial cycles coincide. (A cycle is said to be *nontrivial* if its length is ≥ 2.) □

The collection of lengths of nontrivial cycles of $\sigma \in S(\infty)$ forms a partition without 1s, which will be called the *cycle structure* of σ.

Theorem 3.15 (Integral representation of characters) *There is a bijective correspondence $\chi \leftrightarrow P$ between the characters of the group $S(\infty)$ and the probability Borel measures on the Thoma simplex Ω, determined by formula*

$$\chi(\sigma) = \int_{\Omega} p^{\circ}_{\rho}(\omega)P(d\omega), \qquad \sigma \in S(\infty), \tag{3.17}$$

where partition ρ stands for the cycle structure of σ.

Equivalence of Theorems 3.12 and 3.15 This is an easy consequence of the definitions and Frobenius' formula. However, because of the importance of the fact, we present a detailed argument.

By Proposition 3.5, the characters admit an integral representation parallel to (3.15); we only have to check that its exact form is given by (3.17). Obviously, it suffices to do this for the extreme elements. Then we have to prove the following assertion: given $\omega \in \Omega$, the values of the extreme character χ^{ω} corresponding to the extreme coherent system $\{M^{(n,\omega)}\}_{n\geq 0}$ are given by formula

$$\chi^{\omega}(\sigma) = p^{\circ}_{\rho}(\omega), \qquad \sigma \in S(\infty),$$

which is the particular case of (3.17) when P is the delta-measure at ω.

Choose an arbitrary n, so large that σ is contained in the subgroup $S(n) \subset S(\infty)$, and let partition $\widetilde{\rho} \vdash n$ describe the cycle structure of σ viewed as an element of $S(n)$; denote this element by σ_n. Observe that $\widetilde{\rho}$ coincides with ρ within a few parts equal to 1. By the very definition (see (3.8)), the function $p^{\circ}_{1}(\cdot)$ is identically equal to 1. Therefore,

$$p^{\circ}_{\rho}(\omega) = p^{\circ}_{\widetilde{\rho}}(\omega).$$

Consequently, the desired equality can be equivalently written as

$$\chi^{\omega}_{n}(\sigma_n) = p^{\circ}_{\widetilde{\rho}}(\omega).$$

Next, by the definition of the correspondence between characters and coherent systems, the left-hand side is

$$\chi^{\omega}_{n}(\sigma_n) = \sum_{\lambda \in \mathbb{Y}_n} M^{(n,\omega)}(\lambda) \frac{\chi^{\lambda}_{\widetilde{\rho}}}{\dim \lambda}.$$

But

$$\frac{M^{(n,\omega)}(\lambda)}{\dim \lambda} = s^{\circ}_{\lambda}(\omega).$$

Therefore, we finally reduce the desired relation to

$$p^{\circ}_{\widetilde{\rho}}(\omega) = \sum_{\lambda \in \mathbb{Y}_n} \chi^{\lambda}_{\widetilde{\rho}} s^{\circ}_{\lambda}(\omega),$$

which in turn immediately follows from Frobenius' formula, because $f \mapsto f^\circ$ is an algebra morphism. □

3.5 Exercises

Exercise 3.1 Prove the following properties of the Young graph:

(a) Given a vertex $\lambda \in \mathbb{Y}_n$, the number of adjacent vertices on the next level $n+1$ is equal to the number of those on the preceding level $n-1$ plus 1.

(b) Let κ and λ be two distinct vertices on the same level $\mathbb{Y}_n$, $n \geq 2$. Then there exists at most one vertex $\mu \in \mathbb{Y}_{n-1}$ adjacent to both κ and λ, and at most one vertex $\nu \in \mathbb{Y}_{n+1}$ with the same property. Moreover, μ exists if and only if ν exists.

Exercise 3.2 Prove that a Hausdorff topology in a linear vector space defined by a countable set of semi-norms is always metrizable. Conclude that the topology of pointwise convergence in the space of functions on a countable set is metrizable.

Exercise 3.3 (a) Show that Thoma's simplex with the topology induced from $[0, 1]^\infty \times [0, 1]^\infty \subset \mathbb{R}^\infty \times \mathbb{R}^\infty$ is a compact metrizable topological space.

(b) Show that the subset

$$\Omega_0 = \left\{ (\alpha, \beta) \in \Omega \mid \sum_{i \geq 1} (\alpha_i + \beta_i) = 1 \right\}$$

is dense in Ω. Conclude that the function $\gamma = 1 - \sum_{i \geq 1} (\alpha_i + \beta_i)$ on Ω is not continuous.

Exercise 3.4 Prove formula (3.14).

Exercise 3.5 Show that the statement of the Stone–Weierstrass theorem (Theorem 3.10) becomes false if one removes the assumption $1 \in A$.

3.6 Notes

The notions discussed here originated in the works of Thoma [117] and Vershik and Kerov [121], [122], [124], [125], [126], [63]. In the papers of Vershik and Kerov, the entries of the stochastic matrices Λ^n_{n-1} are called *cotransition probabilities*. In combinatorics, the Young graph is often called the *Young lattice*. The properties of the Young graph $\mathbb{Y}$ indicated in Exercise 3.1 are related to the fact that $\mathbb{Y}$ is a *differential poset* (see comments in Stanley [113, p. 499]).

4

Extreme Characters and Thoma's Theorem

4.1 Thoma's Theorem

Recall that a point of a convex set is said to be *extreme* if it cannot be represented as a nontrivial convex combination of two distinct points of the same set. If a convex set consists of all probability measures on a space, then it is clear that the extreme points are precisely the delta-measures.

Observe now that the bijection stated in Theorem 3.12 is an isomorphism of convex sets. Therefore, the theorem implies the following corollary:

Corollary 4.1 *The extreme coherent systems of distributions on the Young graph are parameterized by points $\omega \in \Omega$. Given ω, the corresponding extreme coherent system $\{M^{(n,\omega)}\}_{n\geq 0}$ is defined by*

$$M^{(n,\omega)}(\lambda) = \dim \lambda \cdot s_\lambda^\circ(\omega), \qquad n = 0, 1, 2, \ldots, \qquad \lambda \in \mathbb{Y}_n. \tag{4.1}$$

Equivalently, the extreme nonnegative normalized harmonic functions on $\mathbb{Y}$ are exactly the functions of the form

$$\psi^{(\omega)}(\lambda) = s_\lambda^\circ(\omega), \qquad \omega \in \Omega, \ldots, \qquad \lambda \in \mathbb{Y}. \tag{4.2}$$

Likewise, Theorem 3.15 implies the following corollary:

Corollary 4.2 (Thoma's theorem) *The extreme characters of the group $S(\infty)$ are parameterized by points of the Thoma simplex Ω. Given $\omega = (\alpha, \beta) \in \Omega$, the values of the corresponding extreme character χ^ω on elements $\sigma \in S(\infty)$ are given by the formula*

$$\chi^\omega(\sigma) = p_\rho^\circ(\omega), \tag{4.3}$$

where ρ is the partition without parts equal to 1, representing the cycle structure of σ.

In more detail, write ρ in the so-called multiplicative form $\rho = (2^{m_2}3^{m_3}\dots)$, meaning that σ has m_2 cycles of length 2, m_3 cycles of length 3, etc. Then

$$\chi^{\omega}(\sigma) = \prod_{k=2}^{\infty}\left(\sum_{i=1}^{\infty}\alpha_i^k + (-1)^{k-1}\sum_{i=1}^{\infty}\beta_i^k\right)^{m_k}. \tag{4.4}$$

The product in the right-hand side of (4.4) is actually finite because $m_k = 0$ for all k large enough. Let us emphasize that the product starts from $k = 2$ and not $k = 1$, because we ignore trivial cycles of length 1. (In formula (4.3), however, we could painlessly include in the index of the p° function any number of 1s, because $p_1^{\circ} \equiv 1$ by the very definition; we just did this in the above proof of equivalence of Theorems 3.12 and 3.15 when we replaced ρ by $\widetilde{\rho}$.)

Given an extreme character $\chi = \chi^{\omega}$, we call the coordinates α_i and β_i of the point $\omega \in \Omega$ the *Thoma parameters* of χ. If we adopt the viewpoint that the extreme characters χ^{ω} of $S(\infty)$ are analogs of the irreducible characters χ^{λ} of the finite symmetric groups $S(n)$, then it is natural to look for a connection between points $\omega \in \Omega$ assembling the Thoma parameters, and Young diagrams λ. Such a connection does exist; see Chapter 6. In particular, Theorem 6.16 explains the *asymptotic meaning* of the Thoma parameters.

4.2 Multiplicativity

Denote by $\mathrm{Cl}(S(\infty))$ the set of conjugacy classes in the group $S(\infty)$. As was explained above, the conjugacy classes can be parameterized by partitions ρ without parts equal to 1. Given two such partitions, ρ' and ρ'', the disjoint union $\rho' \sqcup \rho''$ of their parts is another partition of the same kind. This operation is commutative and associative, and it equips $\mathrm{Cl}(S(\infty))$ with a structure of commutative semigroup with a unit element (the one corresponding to the empty partition, or the class of the unit element in $S(\infty)$).

All characters of $S(\infty)$ can be viewed as functions on this semigroup. A remarkable feature of formula (4.4) is that the extreme characters are *multiplicative functions* on the semigroup $\mathrm{Cl}(S(\infty))$. Moreover, this is their characteristic property. The goal of this section is to prove this fact independently of the classification of characters:

Theorem 4.3 *A character of $S(\infty)$ is extreme if and only if it is multiplicative as a function on the semigroup $\mathrm{Cl}(S(\infty))$.*

First of all, let us state a simple but useful proposition which provides one more interpretation of characters (cf. Proposition 3.5):

Proposition 4.4 *There is a bijective correspondence $\chi \leftrightarrow F$ between characters of the group $S(\infty)$ and linear functionals $F : \mathrm{Sym} \to \mathbb{R}$ satisfying the following three conditions:*

(a) $F(s_\lambda) \geq 0$ *for every* $\lambda \in \mathbb{Y}$;
(b) F *vanishes on* $(p_1 - 1)$, *the principal ideal generated by* $p_1 - 1$;
(c) $F(1) = 1$.

The correspondence is determined in the following way. Given χ, let ρ, as above, range over the set of partitions without parts equal to 1, and let $\chi(\rho)$ mean the value of χ on the corresponding conjugacy class. Then, for every $m = 0, 1, 2, \ldots$,

$$F(p_1^m p_\rho) = \chi(\rho). \tag{4.5}$$

Equivalently, denoting by $\{M^{(n)}\}$ the coherent system of distributions corresponding to χ,

$$F(s_\lambda) = \frac{M^{(n)}(\lambda)}{\dim \lambda}, \qquad \lambda \in \mathbb{Y}, \quad n := |\lambda|. \tag{4.6}$$

The proof is left as an exercise (see Exercise 4.1).

We also need the following classical result.

Theorem 4.5 (Choquet's theorem; see, e.g., Phelps [101]) *Assume that X is a metrizable compact convex set in a locally convex topological space, and let x_0 be an element of X. Then there exists a (Borel) probability measure P on X supported by the set of extreme points $E(X)$ of X, which represents x_0. In other words, for any continuous linear functional f*

$$f(x_0) = \int_{E(X)} f(x) P(dx).$$

($E(X)$ is a G_δ-set; hence it inherits the Borel structure from X.)

Proof of Theorem 4.3 We apply Proposition 4.4 to switch from the language of characters to that of functionals on Sym. Let X stand for the set of functionals F satisfying the three conditions of that proposition. It is a convex set, and the correspondence $\chi \leftrightarrow F$ obviously preserves the convex structure. Next, as is seen from (4.5), χ is a multiplicative function on the semigroup $\mathrm{Cl}(S(\infty))$ if and only if the corresponding functional F is multiplicative; that is to say, $F(fg) = F(f)F(g)$ for any $f, g \in \mathrm{Sym}$. Therefore, the assertion of Theorem 4.3 is equivalent to the following: a functional $F \in X$ is extreme if and only if it is multiplicative. Now we will prove the latter assertion.

Step 1. Let Sym_+ be the cone in Sym formed by linear combinations of the basis elements s_λ with nonnegative coefficients. Observe that this cone is closed under multiplication, because the coefficients of decomposition of $s_\lambda s_\mu$ in the basis of Schur functions are nonnegative (see Exercise 2.13(b)).

Step 2. For any $\lambda \in \mathbb{Y}$ one has

$$p_1^n - \dim\lambda \cdot s_\lambda \in \mathrm{Sym}_+, \qquad n := |\lambda|.$$

Indeed, this follows from the equality (see Exercise 2.11)

$$p_1^n = \sum_{\lambda \in \mathbb{Y}_n} \dim\lambda \, s_\lambda.$$

Step 3. Let $F \in X$ and $f \in \mathrm{Sym}_+$ be such that $F(f) > 0$. Assign to f another functional defined by

$$F_f(g) = \frac{F(fg)}{F(f)}, \qquad g \in \mathrm{Sym}$$

(the definition makes sense because $F(f) \neq 0$). We claim that F_f also belongs to X. Indeed, property (a) follows from Step 1, and properties (b) and (c) are obvious.

Step 4. Assume $F \in X$ is extreme and show that it is multiplicative. It suffices to prove that for any nonempty diagram μ,

$$F(s_\mu g) = F(s_\mu)F(g), \qquad g \in \mathrm{Sym}. \tag{4.7}$$

There are two possible cases: either $F(s_\mu) = 0$ or $F(s_\mu) > 0$.

In the first case, (4.7) is equivalent to saying that $F(s_\mu s_\lambda) = 0$ for every $\lambda \in \mathbb{Y}$. But we have, setting $n = |\lambda|$,

$$0 \le F(s_\mu s_\lambda) \le F(s_\mu p_1^n) = F(s_\mu) = 0,$$

where the first inequality follows from step 1, the second one follows from step 2, and the equality holds by virtue of property (b). Therefore, $F(s_\mu s_\lambda) = 0$, as desired.

In the second case we may form the functional F_{s_μ}, and then (4.7) is equivalent to saying that it coincides with F. Let $m = |\mu|$ and set

$$f_1 = \frac{1}{2}\dim\mu \, s_\mu, \qquad f_2 = p_1^m - f_1.$$

Observe that F is strictly positive both on f_1 and on f_2, so that both F_{f_1} and F_{f_2} exist. On the other hand, for any $g \in \mathrm{Sym}$ one has

$$F(g) = F(p_1^m g) = F(f_1 g) + F(f_2 g), \qquad F(f_1) + F(f_2) = 1,$$

which entails that F is a convex combination of F_{f_1} and F_{f_2} with strictly positive coefficients:

$$F = F(f_1)F_{f_1} + F(f_2)F_{f_2}.$$

Since F is extreme, we conclude that $F_{f_1} = F$, as desired.

Step 5. Conversely, assume that $F \in X$ is multiplicative and let us show that it is extreme.

Observe that X satisfies the assumptions of Choquet's theorem (Theorem 4.5). Indeed, we may regard X as a subset of the vector space dual to Sym and equipped with the topology of simple convergence.

Let P be the probability measure on $E(X)$ representing F in accordance with Choquet's theorem. This implies that

$$F(f) = \int_{G\in E(X)} G(f)P(dG), \qquad f \in \mathrm{Sym}. \tag{4.8}$$

We are going to prove that P is actually the delta-measure at a point of $E(X)$, which just means that F is extreme.

Write ξ_f for the function $G \to G(f)$ viewed as a random variable defined on the probability space $(E(X), P)$. Equality (4.8) says that the mean of ξ_f equals $F(f)$.

By virtue of step 4, every $G \in E(X)$ is multiplicative. On the other hand, F is multiplicative by the assumption. Using this we get from (4.8)

$$\begin{aligned}\left(\int_{G\in E(X)} G(f)P(dG)\right)^2 = (F(f))^2 = F(f^2) &= \int_{G\in E(X)} G(f^2)P(dG)\\ &= \int_{G\in E(X)} (G(f))^2 P(dG).\end{aligned}$$

Comparing the leftmost and the rightmost expressions we conclude that ξ_f has zero variance. Hence $G(f) = F(f)$ for all $G \in E(X)$ outside a P-null subset depending on f.

It follows that $G(s_\lambda) = F(s_\lambda)$ for all $\lambda \in \mathbb{Y}$ and all $G \in E(X)$ outside a P-null subset, which is only possible when P is a delta-measure.

This completes the proof. □

4.3 Exercises

Exercise 4.1 Prove Proposition 4.4. (Hint: use the special case of Exercise 2.9 corresponding to $r = 1$.)

Exercise 4.2 (a) Assume that:
(i) A is a commutative unital algebra A over $\mathbb{R}$ and $K \subset A$ is a convex cone defining a partial order in A (below we write $a \geq b$ if $a - b \in K$);
(ii) the cone is generating (that is, $K - K = A$) and stable under multiplication (that is, $K \cdot K \in K$);
(iii) the unit element $1 \in A$ belongs to the cone (that is, $1 \geq 0$) and for any $a \in K$ there exists an $\varepsilon > 0$ such that $\varepsilon \cdot a \leq 1$;
(iv) the cone K is generated by a countable set of elements.

Next, let A' be the dual space to A and $K_1' \subset A'$ be the set of all linear functionals $F : A \to \mathbb{R}$ which are nonnegative on K and normalized by $F(1) = 1$. Observe that K_1' is a convex set.

Under these assumptions, prove that a functional $F \in K_1'$ is an extreme point of K_1' if and only if F is multiplicative; that is, $F(ab) = F(a)F(b)$ for any $a, b \in A$. (Hint: adapt the proof of Theorem 4.3.)

The above claim is the so-called *ring theorem* due to Vershik and Kerov [69], [63]; see also a more detailed exposition in Gnedin and Olshanski [45].

(b) Show that Theorem 4.3 is a special case of the ring theorem with A being the quotient of the algebra Sym of symmetric functions by the principal ideal $(p_1 - 1)$, and K being the image of the convex cone $\mathrm{Sym}_+ \subset \mathrm{Sym}$ generated by the Schur functions under the projection $\mathrm{Sym} \to \mathrm{Sym}/(p_1 - 1)$.

Exercise 4.3 Let χ be a class function on $S(\infty)$, that is, χ is constant on conjugacy classes. Prove that χ is multiplicative as a function on the semigroup $\mathrm{Cl}(S(\infty))$ of conjugacy classes (= cycle structures) if and only if the following analog of the functional equation holds (cf. Exercise 1.5):

$$\lim_{n\to\infty} \frac{1}{n!} \sum_{h \in S(n)} \chi(g_1 h g_2 h^{-1}) = \chi(g_1)\chi(g_2), \qquad g_1, g_2 \in S(\infty).$$

Equivalently, as $n \to \infty$, the structure constants in the algebra of conjugacy classes of the group $S(n)$ tend to the structure constants of a semigroup. Details can be found in Olshanski [89].

Exercise 4.4 Without referring to Thoma's theorem, prove that the pointwise product of two extreme characters is extreme. Thus, the set of extreme characters is a semigroup. Describe the product in terms of Thoma's parameters (α, β).

Exercise 4.5 Let, as above, $\mathrm{Sym}_+ \subset \mathrm{Sym}$ denote the convex cone generated by the Schur functions and let Sym_+' be the dual cone in the dual space to Sym. By virtue of Theorem 4.3 and Proposition 4.4, extreme characters of $S(\infty)$ correspond to those multiplicative linear functionals $F : \mathrm{Sym} \to \mathbb{R}$

which are contained in the cone Sym'_+ and satisfy $F(p_1) = 1$. Then Thoma's theorem (Corollary 4.2 to Theorem 3.12) is equivalent to the statement that these functionals are precisely those of the form

$$F_\omega : f \mapsto f^\circ(\omega), \qquad f \in \mathrm{Sym}, \quad \omega \in \Omega$$

(recall that $f \mapsto f^\circ$ is defined in Proposition 3.9).

Here we sketch a proof of the fact that F_ω possesses the required properties for any $\omega \in \Omega$, which is independent of the proof of Theorem 3.12 presented in Chapter 6. However, the argument below does not prove the completeness of the list $\{F_\omega : \omega \in \Omega\}$. Thus, this argument will only show that the functions χ^ω on $S(\infty)$ are extreme characters but not the fact that they exhaust all such ones.

(a) The idea is to use the comultiplication map Δ (see Exercise 2.16) to define an operation in Sym'_+, which produces a variety of elements of Sym'_+ starting from two "elementary" ones.

Let F_1 and F_2 be two elements of Sym'_+ and let u and v be two nonnegative real numbers with sum 1. Define the mixing of F_1 and F_2 in proportion $u : v$ as follows: given a homogeneous element $f \in \mathrm{Sym}$, take a finite expansion

$$\Delta(f) = \sum_i g_i^{(1)} \otimes g_i^{(2)}$$

with homogeneous elements $g_i^{(1)}$, $g_i^{(2)}$ and set

$$F(f) = \sum_i u^{\deg g_i^{(1)}} v^{\deg g_2^{(2)}} F_1(g_i^{(1)}) F_2(g_i^{(2)}).$$

Show that this definition does not depend on the choice of the expansion above. (Hint: let $r_u : \mathrm{Sym} \to \mathrm{Sym}$ be the algebra endomorphism reducing to multiplication by u^k on the homogeneous component $\mathrm{Sym}^k \subset \mathrm{Sym}$. Observe that F is obtained from $F_1 \otimes F_2$ by superposition of two maps: first, the dual to $r_u \otimes r_v$, and, next, the dual to Δ.)

(b) Show that F, as defined above, belongs to Sym'_+. (Hint: use the non-negativity of the structure constants of Δ in the basis $\{s_\lambda\}$; see Exercise 2.16, item (d).)

(c) Show that F is multiplicative if F_1 and F_2 are multiplicative. (Hint: use the fact that Δ, r_u, and r_v are multiplicative.)

(d) Show that $F(p_1) = 1$ if $F_1(p_1) = F_2(p_1) = 1$.

(e) Generalize the construction to the case when there are $n = 2, 3, \dots$ functionals $F_1, \dots, F_n$ and n nonnegative reals number $u_1, \dots, u_n$ with sum 1.

(f) Let F and G be the multiplicative functionals corresponding to parameters (α, β) with $\alpha_1 = 1$ and $\beta_1 = 1$, respectively (the remaining coordinates in each of the two cases are automatically equal to 0). Show that both F and G are in Sym'_+. Let us call these functionals *elementary* ones.

(g) Check that F_ω belongs to Sym'_+ under the assumption that there are only finitely many nonzero coordinates α_i and β_j in ω and their total sum equals 1. (Hint: apply (e) taking as the numbers $u_1, u_2, \ldots$ the collection of nonzero coordinates in α and β and choosing as $F_1, F_2, \ldots$ one of the two elementary functionals from (f).)

(h) Finish the proof using a continuity argument.

4.4 Notes

The description of extreme characters given in Corollary 4.2 is due to Thoma [117]. The idea to establish first the integral representation theorem for general characters and then obtain Thoma's theorem as a corollary was realized in the paper [64] by Kerov, Okounkov and Olshanski.

The multiplicativity property of the extreme characters of $S(\infty)$ was discovered by Thoma, but the proof that we give is due to Kerov and Vershik [69]. Some details (omitted in that paper) were communicated to one of us by Kerov and were later included in the paper of Gnedin–Olshanski [45].

The multiplicativity property for extreme characters and extreme spherical functions of "big groups" is a general phenomenon, which can be explained in various ways. See the survey paper Olshanski [89].

The construction described in Exercise 4.5 (generation of extreme characters by making use of the comultiplication in Sym) is due to Kerov [59]; his idea was exploited in [45].

5

A Toy Model (the Pascal Graph) and de Finetti's Theorem

Instead of going into the proof of Theorem 3.12 right away, we start with a similar assertion for a much simpler object – the *Pascal graph* $\mathbb{P}$.

By definition, $\mathbb{P}$ is a $\mathbb{Z}_{\geq 0}$-graded graph whose nth level consists of all pairs $(k, l) \in \mathbb{Z}^2_{\geq 0}$ subject to the condition $k + l = n$. The edges of $\mathbb{P}$ correspond to shifts of one of the coordinates by 1. Note that this is a subgraph of $\mathbb{Y}$ made of all hook Young diagrams (that is, diagrams with exactly one diagonal box); to (k, l) one assigns the partition $(k + 1, 1^l)$. The symmetry $(k, l) \to (l, k)$ of $\mathbb{P}$ agrees with the symmetry of the Young graph induced by transposition of diagrams. Note that the nth level of $\mathbb{P}$ is embedded into the $(n + 1)$st level of the Young graph.

We will use the language of harmonic functions (cf. Definition 3.6). These are real-valued functions $\psi(k, l)$ on $\mathbb{Z}^2_{\geq 0}$ satisfying

$$\psi(k, l) = \psi(k + 1, l) + \psi(k, l + 1), \qquad k, l \geq 0.$$

As above, we are interested in nonnegative harmonic functions normalized by the condition $\psi(0, 0) = 1$.

Theorem 5.1 *The nonnegative harmonic normalized functions ψ on the Pascal graph $\mathbb{P}$ are in one-to-one correspondence with the probability measures P on $[0, 1]$; the bijection has the form*

$$\psi(k, l) = \int_0^1 \alpha^k (1 - \alpha)^l P(d\alpha). \tag{5.1}$$

In particular, the extreme functions are precisely those of the form $\psi_\alpha(k, l) = \alpha^k (1 - \alpha)^l$, $\alpha \in [0, 1]$.

Before proving Theorem 5.1 we will discuss its equivalent formulation. Consider the infinite product space

$$\{0, 1\}^\infty = \{(x_1, x_2, \dots) \mid x_i = 0 \text{ or } 1\}.$$

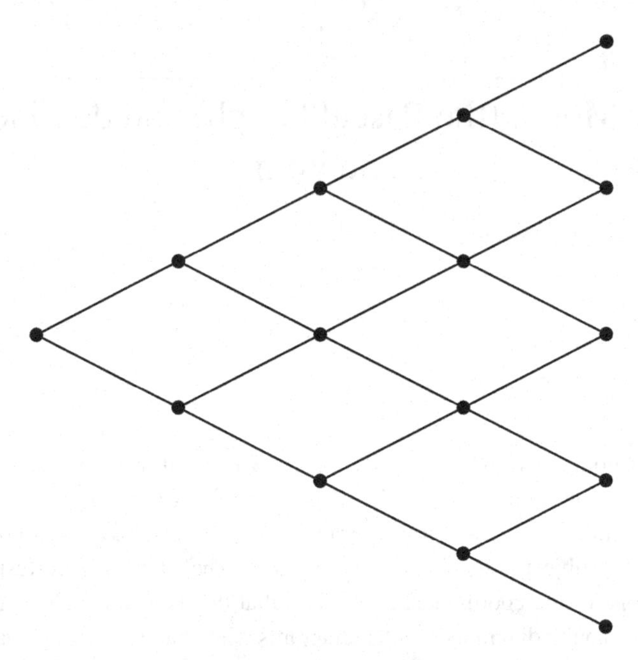

Figure 5.1 The Pascal graph.

Denote by ν_α, where $\alpha \in [0, 1]$, the probability measure on $\{0, 1\}$ assigning weights α and $1 - \alpha$ to the points 1 and 0, respectively, and let $\nu_\alpha^{\otimes\infty}$ denote the product of infinitely many copies of ν_α. This is a probability measure on $\{0, 1\}^\infty$ called the (stationary) *Bernoulli measure* with parameter α. Note that the group $S(\infty)$ acts on the space $\{0, 1\}^\infty$ by permuting the coordinates x_i, and every Bernoulli measure $\nu_\alpha^{\otimes\infty}$ is invariant with respect to this action.

Theorem 5.2 (de Finetti's theorem) *The $S(\infty)$-invariant probability measures ν on $\{0, 1\}^\infty$ are in one-to-one correspondence with probability measures P on $[0, 1]$; the bijection has the form*

$$\nu = \int_0^1 \nu_\alpha^{\otimes\infty} P(d\alpha).$$

That is, invariant probability measures are precisely mixtures of stationary Bernoulli measures. In particular, the latter measures are the extreme points in the convex set of all invariant probability measures.

Proposition 5.3 *The convex sets $\{\psi\}$ and $\{\nu\}$ appearing in Theorem 5.1 and Theorem 5.2, respectively, are naturally isomorphic, and under this isomorphism, $\psi_\alpha \leftrightarrow \nu_\alpha^{\otimes\infty}$. Thus, the two theorems are equivalent.*

Proof Given $(k, l) \in \mathbb{Z}^2_{\geq 0}$, let $C_{k,l}$ denote the cylindric subset in $\{0, 1\}^\infty$ formed by all sequences $(x_i) \in \{0, 1\}^\infty$ such that $x_1 = \cdots = x_k = 1$ and $x_{k+1} = \cdots = x_{k+l} = 0$. Note that for every fixed $n = 1, 2, \ldots$, the space $\{0, 1\}$ can be represented as a disjoint union of 2^n cylindric subsets,

$$\{0, 1\}^\infty = \bigsqcup_{k+l=n} \bigsqcup_{\sigma \in S(n)/S(k)\times S(l)} \sigma(C_{k,l}),$$

where, given (k, l) with $k + l = n$, σ ranges over a set of representatives in $S(n)$ of the cosets modulo $S(k) \times S(l)$.

Let ν be an invariant probability measure on $\{0, 1\}^\infty$. Then we set $\psi(k, l) = \nu(C_{k,l})$, which also equals $\nu(\sigma(C_{k,l}))$ for any $\sigma \in S(k + l)$. Observe that $C_{k,l}$ is the disjoint union of $C_{k,l+1}$ and a shift of $C_{k+1,l}$ under an appropriate permutation $\sigma \in S(k + l + 1)$. This implies that ψ is harmonic. Clearly, ψ is nonnegative. It is normalized, because $C_{0,0}$ coincides with the whole space $\{0, 1\}^\infty$.

Conversely, let ψ be harmonic, nonnegative, and normalized. We define nonnegative function ν on cylindric sets by setting

$$\nu(\sigma(C_{k,l})) = \psi(k, l), \qquad (k, l) \in \mathbb{Z}^2_{\geq 0}, \quad \sigma \in S(k + l).$$

Then harmonicity of ψ implies that ν can be interpreted as a consistent system of probability measures on the finite product spaces $\{0, 1\}^n$, $n = 1, 2, 3, \ldots$. By the Kolmogorov extension theorem (see Theorem 7.9, below) this system determines a probability measure on $\{0, 1\}^\infty$. It is $S(\infty)$-invariant, because for every n, the corresponding measure on $\{0, 1\}^n$ is $S(n)$-invariant.

The last assertion, concerning the correspondence $\psi_\alpha \leftrightarrow \nu_\alpha^{\otimes\infty}$, is readily verified. □

By virtue of Proposition 5.3, Theorem 5.1 follows from de Finetti's theorem. However, we will prove Theorem 5.1 independently, in two different ways.

First proof of Theorem 5.1 Arguing as in the case of the Young graph (see the end of Chapter 4), using Choquet's theorem we prove that any nonnegative normalized harmonic function is a convex combination of the extreme ones.

Lemma 5.4 *Extreme nonnegative normalized harmonic functions on the Pascal graph are in one-to-one correspondence with algebra homomorphisms* $F : \mathbb{R}[x, y] \to \mathbb{R}$ *such that* $F(x + y) = 1$ *and*

$$F(x^k y^l) \geq 0, \qquad k, l = 0, 1, 2, \ldots.$$

The bijection is given by $\psi(k, l) = F(x^k y^l)$.

Proof The argument is exactly the same as for the Young graph; see the proof of Theorem 4.3. The role of p_1 and Pieri's rule $p_1 s_\lambda = \sum_{\nu:\nu\searrow\lambda} s_\nu$ is played by $x + y$ and the formula

$$(x+y)x^k y^l = x^{k+1}y^l + x^k y^{l+1}.$$

One can also apply the ring theorem (see Exercise 4.2, above). □

From this lemma it is easily seen that the extreme normalized nonnegative harmonic functions are precisely those of the form $\psi(k,l) = \alpha^k(1-\alpha)^l$, where $\alpha \in [0,1]$ (here parameter α arises as the image of x under F). Therefore, arbitrary nonnegative normalized harmonic functions are given by formula (5.1).

It remains to prove that the correspondence $P \mapsto \psi$ determined by (5.1) is injective. That is, a harmonic function ψ uniquely determines the measure P on $[0,1]$ that represents it. This immediately follows from the fact that the linear span of functions $\{\alpha^k(1-\alpha)^l \mid k,l \geq 0\}$ is dense in $C([0,1])$. □

We were lucky that for the Pascal graph we could classify all needed algebra homomorphisms explicitly. This is a much harder problem for $\mathbb{Y}$. That is why we will give a different proof of de Finetti's theorem that will be generalized to the case of $\mathbb{Y}$ in Chapter 6.

Second proof of Theorem 5.1 Let P be a probability measure on $[0,1]$. Then it is easily seen that the function ψ defined by (5.1) is a nonnegative harmonic function normalized at the root vertex $(0,0)$. Moreover, as was pointed out above, the map $P \mapsto \psi$ is injective. The nontrivial part of the theorem consists in proving that the map is surjective. We need two preliminary statements.

Lemma 5.5 *Let K be a compact topological space. The space of Borel probability measures on K is compact in the weak topology.*

Proof The unit ball in the dual space to a normed space is compact in the weak-star topology (this is classical Banach–Alaoglu theorem; see, e.g., Reed and Simon [105, Theorem IV.21]). Take the normed space to be $C(K)$, the space of continuous functions on K with the sup-norm. Its dual consists of all signed measures on K, and the weak-star topology is the weak topology on measures (this is Riesz's theorem; see, e.g., Bogachev [5, Theorem 7.10.4]). Finally, observe that probability measures form a closed subset of the unit ball. □

Let $\dim((k_0,l_0),(k,l))$ denote the number of paths in the Pascal graph going from (k_0,l_0) to (k,l) and increasing the level number by one on every step. Clearly,

$$\dim((k_0, l_0), (k, l)) = \binom{k + l - k_0 - l_0}{k - k_0} = \frac{(k + l - k_0 - l_0)!}{(k - k_0)!(l - l_0)!}.$$

By the very definition, the quantity $\dim((k_0, l_0), (k, l))$ vanishes unless $k_0 \leq k$ and $l_0 \leq l$; note the above formula agrees with this fact.

Lemma 5.6 *We have*

$$\frac{\dim((k_0, l_0), (k, l))}{\dim((0, 0), (k, l))} = \left(\frac{k}{k + l}\right)^{k_0} \left(\frac{l}{k + l}\right)^{l_0} + O\left(\frac{1}{k + l}\right),$$

where the estimate is uniform on $\{(k, l) \in \mathbb{Z}^2_{\geq 0} \mid k + l > k_0 + l_0\}$.

Proof Denote $n^{\downarrow m} = n(n - 1) \cdots (n - m + 1)$. Observe that if m is fixed then $n^{\downarrow m}$ is a polynomial in n with highest degree term n^m. The needed ratio of numbers of paths equals

$$\frac{k^{\downarrow k_0} l^{\downarrow l_0}}{(k + l)^{\downarrow (k_0 + l_0)}} = \frac{(k^{k_0} + \cdots)(l^{l_0} + \cdots)}{(k + l)^{k_0 + l_0} + \cdots},$$

where dots mean lower degree terms, which implies the claim of the lemma. □

We continue the proof of Theorem 5.1. In what follows we use the notation

$$\dim(k, l) := \dim((0, 0), (k, l)) = \frac{(k + l)!}{k!\,l!}.$$

Let ψ be a nonnegative normalized harmonic function on the Pascal graph. Our aim is construct a probability measure P such that (5.1) holds true. Observe that for any $n = 0, 1, 2, \ldots$, the numbers

$$M^{(n)}(k, l) = \dim(k, l) \cdot \psi(k, l) = \frac{(k + l)!}{k!\,l!} \cdot \psi(k, l), \qquad k + l = n,$$

determine a probability distribution on the nth level of the graph. (Indeed, summing up these weights over the nth level is equivalent to computing the total measure of $\{0, 1\}^\infty$ by adding up measures of cylindric subsets with fixed first n coordinates.)

The $M^{(n)}$ are analogs of coherent systems of distributions on $\mathbb{Y}$ considered in Chapter 3.

The harmonicity condition and Lemma 5.6 imply

$$\begin{aligned}
\psi(k_0, l_0) &= \sum_{k+l=n} \dim((k_0, l_0), (k, l)) \cdot \psi(k, l) \\
&= \sum_{k+l=n} \frac{\dim((k_0, l_0), (k, l))}{\dim(k, l)} M^{(n)}(k, l)
\end{aligned}$$

$$= \sum_{k+l=n} \left(\frac{k}{k+l}\right)^{k_0} \left(\frac{l}{k+l}\right)^{l_0} M^{(n)}(k,l) + O(n^{-1}).$$

For an arbitrary point $p \in [0, 1]$, let $\langle p \rangle$ denote the Dirac delta-measure concentrated at p. Consider the sequence of probability measures

$$P^{(n)} = \sum_{k+l=n} M^{(n)}(k,l) \left\langle \frac{k}{k+l} \right\rangle$$

on [0,1]. By Lemma 5.5, it has a convergent subsequence. Denoting the limit measure by P, we obtain

$$\psi(k_0, l_0) = \int_0^1 p^{k_0}(1-p)^{l_0} P(dp)$$

for any $(k_0, l_0) \in \mathbb{Z}^2_{\geq 0}$. This completes the proof of the theorem. □

5.1 Exercises

Exercise 5.1 (Pascal pyramids) A natural generalization of the Pascal graph can be obtained by replacing $\mathbb{Z}^2_{\geq 0}$ by $\mathbb{Z}^N_{\geq 0}$, where $N = 3, 4, \ldots$. Thus, the vertices are the points of $\mathbb{Z}^N_{\geq 0}$, and the edges correspond to shifts of one of the coordinates by 1. Generalize Theorem 5.1 to this graph, and find an appropriate version of Theorem 5.2.

Exercise 5.2 Extreme invariant measures that appear in Theorem 5.2 can be characterized as follows. Consider a general setting: $\mathfrak{X}$ is a measurable space (that is, a set with a distinguished sigma-algebra of subsets) and G is a group acting on $\mathfrak{X}$ by measurable transformations (in the context of Theorem 5.2, $\mathfrak{X} = \{0, 1\}^\infty$ and $G = S(\infty)$). Consider the convex set of all G-invariant probability measures on $\mathfrak{X}$. Prove that for a measure ν from this set the following conditions are equivalent:

(1) ν is extreme.

(2) Any invariant modulo 0 measurable set in $\mathfrak{X}$ has ν-measure 0 or 1.[1]

(3) The subspace of G-invariant vectors in $L^2(\mathfrak{X}, \nu)$ is one-dimensional; that is, it consists of the constant functions.

Condition (2) is usually taken as the definition of *ergodic* measures. If G is a countable group, then the words "modulo 0" in condition (2) can be omitted.

For more detail, see Phelps [101, Section 10].

[1] A set $A \subset \mathfrak{X}$ is called *invariant modulo 0* if for any $g \in G$, the symmetric difference between A and $g(A)$ has ν-measure 0.

Exercise 5.3 Prove, without using de Finetti's theorem, that every $S(\infty)$-invariant probability measure on $\{0, 1\}^\infty$ is also invariant under the larger group of all (not necessarily finite) permutations of the coordinates.

5.2 Notes

A different approach to de Finetti's theorem 5.2 can be found in Feller's textbook [39, Chapter VII]. Feller also explains a close connection between de Finetti's theorem, the Hausdorff moment problem, and classical Bernstein's polynomials. In our view, this connection can be best explained via the Pascal graph.

What we called de Finetti's theorem is actually its simplest version. In its full generality, de Finetti's theorem deals with general product spaces X^∞ (for instance, X is $\mathbb{R}$ or an arbitrary Borel subset of $\mathbb{R}$), and it establishes a bijection between $S(\infty)$-invariant probability measures and arbitrary *random* probability measures on X. See, e.g., Hewitt and Savage [56], Aldous [3].

One way of generalizing Theorem 5.1 is to keep $\mathbb{Z}^2_{\geq 0}$ as the vertex set but equip the edges with (possibly formal) multiplicities, which results in a deformation of the harmonicity equation. Examples can be found in Kerov [63, Chapter 1, Section 2], Gnedin–Pitman [50], Gnedin–Olshanski [46]. The boundary of such graphs substantially depends on the concrete choice of multiplicities.

One of Kerov's examples in [63, Chapter 1, Section 2] is the q-Pascal graph. As explained in Gnedin–Olshanski [47], it is related to a q-version of de Finetti's theorem. This subject was further developed in Gnedin–Olshanski [48], [49].

6

Asymptotics of Relative Dimension in the Young Graph

Here we prove Theorem 3.12 (integral representation of coherent systems) and Theorem 6.16 (the Vershik–Kerov theorem about asymptotics of irreducible characters).

6.1 Relative Dimension and Shifted Schur Polynomials

Definition 6.1 The *relative dimension* in the Young graph is the function $\dim(\mu, \lambda)$ on $\mathbb{Y} \times \mathbb{Y}$ whose value at a pair of two diagrams μ, λ is the number of paths in $\mathbb{Y}$ going from μ to λ and increasing the level number by one at every step. This means that $\dim(\mu, \lambda)$ vanishes unless μ is contained in λ. We also agree that $\dim(\mu, \mu) = 1$.

In particular, $\dim(\mu, \lambda)$ vanishes if $|\mu| > |\lambda|$. Note that if μ is contained in λ then $\dim(\mu, \lambda)$ is a nonzero number equal to the number of standard tableaux of the skew shape λ/μ. Evidently,

$$\dim(\varnothing, \lambda) = \dim \lambda.$$

The importance of the notion of relative dimension will become clear later on.

Proposition 6.2 *For any integers $n > m \geq 0$ and any $\mu \in \mathbb{Y}_m$*

$$p_1^{n-m} \cdot s_\mu = \sum_{\lambda \in \mathbb{Y}_n} \dim(\mu, \lambda) s_\lambda. \tag{6.1}$$

Proof Assume first $n = m + 1$. Then (6.1) means

$$p_1 \cdot s_\mu = \sum_{\lambda:\, \mu \nearrow \lambda} s_\lambda, \tag{6.2}$$

which is a particular case of Proposition 2.17 or Pieri's formula (see Exercise 2.9). For $n \geq m$ the desired result is obtained by iterating (6.12) and using the fact that $\dim(\mu, \lambda)$ equals the number of monotone paths in $\mathbb{Y}$ going from μ to λ. □

The next proposition is a generalization of (1.2).

Proposition 6.3 (Aitken's theorem) *For any two Young diagrams μ and λ, and for any $N \geq \ell(\lambda)$, we have*

$$\frac{\dim(\mu, \lambda)}{(|\lambda| - |\mu|)!} = \det\left[\frac{1}{(\lambda_i - \mu_j - i + j)!}\right]_{i,j=1}^{N}. \tag{6.3}$$

When $\mu = \varnothing$, this reduces to (1.2).

Proof If μ is not contained in λ then both sides of (6.3) vanish. Indeed, the left-hand side vanishes because $\dim(\mu, \lambda) = 0$ by the very definition. As for the right-hand side, because μ is not contained in λ, there exists index k such that $\lambda_k < \mu_k$. Then

$$\lambda_i - \mu_j - i + j < 0 \qquad \text{for all } i, j \in \{1, \dots, N\} \text{ such that } i \geq k \geq j.$$

Since $1/p! = 0$ for $p < 0$, for all such (i, j) the corresponding entries of the matrix in the right-hand side vanish, which implies that the determinant equals 0.

Next, if $\mu = \lambda$ then both sides of (6.3) equal 1. Indeed, for the left-hand side this is evident, and for the right-hand side this holds because the matrix is upper triangular with 1s on the diagonal.

Thus, we may assume $\mu \subset \lambda$, which also implies $|\mu| < |\lambda|$ and $\ell(\mu) \leq \ell(\lambda)$. Fix any $N \geq \ell(\lambda)$; then we also have $N \geq \ell(\mu)$. Therefore, the N-variate Schur polynomials indexed by μ and λ are not vanishing.

Now specialize identity (6.1) to N variables $x_1, \dots, x_N$. Multiplying both sides by a_δ (see (2.1) for the definition), we see that $\dim(\mu, \lambda)$ is equal to the coefficient of $x_1^{\lambda_1+N-1} x_2^{\lambda_2+N-2} \cdots x_N^{\lambda_N}$ in $a_{\mu+\delta} \cdot (x_1 + \cdots + x_N)^{n-m}$, where we set $n = |\lambda|$, $m = |\mu|$. Hence, setting

$$l_i = \lambda_i + N - i, \quad m_i = \mu_i + N - i$$

and noting that $\sum_{i=1}^{N}(l_i - m_i) = n - m$, we get

$$\dim(\mu, \lambda) = \sum_{\sigma \in S(N)} \operatorname{sgn}(\sigma) \binom{n - m}{l_1 - m_{\sigma(1)}, \ \dots, \ l_N - m_{\sigma(N)}}$$

$$= (n - m)! \det\left[\frac{1}{(l_i - m_j)!}\right]_{i,j=1}^{N} = (n - m)! \det\left[\frac{1}{(\lambda_i - i - \mu_j + j)!}\right]_{i,j=1}^{N},$$

which is the desired result. □

As before, we use the notation

$$x^{\downarrow m} = \begin{cases} x(x-1)\cdots(x-m+1), & m = 1, 2, \ldots, \\ 1, & m = 0. \end{cases}$$

Definition 6.4 For any partitions μ, we define the *shifted Schur polynomial* in N variables $x_1, \ldots, x_N$, indexed by μ, by

$$s^*_\mu(x_1, \ldots, x_N) = \begin{cases} \dfrac{\det\left[(x_i + N - i)^{\downarrow(\mu_j+N-j)}\right]_{i,j=1}^N}{\det\left[(x_i + N - i)^{\downarrow(N-j)}\right]_{i,j=1}^N}, & N \geq \ell(\mu), \\ 0, & N < \ell(\mu). \end{cases}$$

This definition is quite similar to that of the conventional Schur polynomial given in Chapter 2. One easily shows that:

- the denominator in the definition of s^*_μ is equal to the shifted Vandermonde determinant $\prod_{1\leq i<j\leq N}((x_i - i) - (x_j - j))$;
- the shifted Schur polynomial $s^*_\mu(x_1, \ldots, x_N)$ is an element of $\mathbb{R}[x_1, \ldots, x_N]$ whose highest homogeneous component is equal to the conventional Schur polynomial $s_\mu(x_1, \ldots, x_N)$;
- $s^*_\mu(x_1, \ldots, x_N, 0) = s^*_\mu(x_1, \ldots, x_N)$. Thus, we can correctly define the value of s^*_μ on infinite sequences $(x_1, x_2, \ldots)$ with finitely many nonzero entries.

In particular, the quantity

$$s^*_\mu(\lambda) := s^*_\mu(\lambda_1, \lambda_2, \ldots)$$

is well defined for $\lambda \in \mathbb{Y}$.

Note that

$$s^*_\varnothing(\lambda) \equiv 1.$$

Proposition 6.5 *For any partitions μ and λ*

$$\frac{\dim(\mu, \lambda)}{\dim \lambda} = \frac{s^*_\mu(\lambda)}{n^{\downarrow m}}, \qquad m = |\mu|, \quad n = |\lambda|.$$

Proof Fix N large enough. Applying Proposition 6.3 together with Exercise 1.1, we get

$$\dim \lambda = n! \det\left[\frac{1}{(\lambda_i - i + j)!}\right]_{i,j=1}^N = n! \frac{\prod_{1\leq i<j\leq N}(\lambda_i - i - \lambda_j + j)}{\prod_{i=1}^N (\lambda_i + N - i)!}.$$

Then, applying Proposition 6.3 again, we have

$$\frac{\dim(\mu,\lambda)}{\dim\lambda} = \frac{1}{n^{\downarrow m}} \det\left[\frac{1}{(\lambda_i - i + \mu_j + j)!}\right]_{i,j=1}^{N} \times \frac{\prod_{i=1}^{N}(\lambda_i + N - i)!}{\prod_{1\le i<j\le N}(\lambda_i - i - \lambda_j + j)}.$$

Since

$$\frac{(\lambda_i + N - i)!}{(\lambda_i - i - \mu_j + j)!} = (\lambda_i + N - i)^{\downarrow(\mu_j + N - j)},$$

we are done. □

6.2 The Algebra of Shifted Symmetric Functions

Observe that the shifted Schur polynomials $s^*_\mu(x_1,\ldots,x_N)$ are symmetric with respect to permutations of shifted variables $x_1 - 1,\ldots,x_N - N$ (thus the name). Similarly to the construction of the algebra Sym of symmetric functions in the beginning of Chapter 2, one can carry out the construction of the algebra Sym* of *shifted* symmetric functions.

More precisely, let Sym^*_N be the subalgebra in $\mathbb{R}[x_1,\ldots,x_N]$ formed by polynomials that are symmetric in shifted variables $x_j - j$, $j = 1,\ldots,N$. For $N' > N$, define the projection map $\pi_{N',N} : \mathrm{Sym}^*_{N'} \to \mathrm{Sym}^*_N$ by setting the variables $x_{N+1},\ldots,x_{N'}$ to 0.

The algebra Sym^*_N has an ascending filtration

$$\mathrm{Sym}^{*\,0}_N \subset \mathrm{Sym}^{*\,1}_N \subset \cdots \subset \mathrm{Sym}^{*\,k}_N \subset \cdots, \qquad \mathrm{Sym}^*_N = \bigcup_{k\ge 0} \mathrm{Sym}^{*\,k}_N,$$

where $\mathrm{Sym}^{*\,k}_N$ is formed by polynomials from Sym^*_N of degree $\le k$. We set

$$\mathrm{Sym}^{*k} = \varprojlim \mathrm{Sym}^{*\,k}_N, \qquad \mathrm{Sym}^* = \bigcup_{k\ge 0} \mathrm{Sym}^{*k}.$$

In other words, an element of Sym* is a sequence $f^* = (f^*_0, f^*_1, \ldots)$ such that for some k

$$f^*_N \in \mathrm{Sym}^{*\,k}_N, \quad \pi_{N',N}(f^*_{N'}) = f^*_N \quad (N' > N), \quad N = 0, 1, 2, \ldots.$$

In particular, the shifted Schur polynomials $s^*_\mu(x_1,\ldots,x_N) \in \mathrm{Sym}^{*\,k}_N$ with $k = |\mu|$ and $N = 0, 1, 2, \ldots$ define an element of Sym* that we call the *shifted Schur function* and denote by the same symbol s^*_μ.

If $f^* = (f^*_0, f^*_1, \ldots)$ is an element of Sym^{*k}, then for every $N = 1, 2, \ldots,$ the degree k homogeneous component f_N of f^*_N is a symmetric polynomial in N variables, and these polynomials are consistent with projections $\pi_{N',N}$. Thus, the sequence $f = (f_0, f_1, \ldots)$ determines a symmetric function, which is called the *highest term* of f^* and denoted as $f = [f^*]$.[1] One easily shows that the operation of taking the highest term is a vector space isomorphism of $\mathrm{Sym}^{*k}/\mathrm{Sym}^{*k-1}$ and Sym^k (see Chapter 2 for the definition of the latter). Since the highest term of the product of two shifted symmetric functions is the product of the highest terms of the factors, Sym coincides with the graded algebra associated to the filtered algebra Sym^*.

One immediate corollary of the above observations is that the shifted Schur functions $\{s^*_\mu\}_{\mu\in\mathbb{Y}}$ form a linear basis in Sym^*. Indeed, $[s^*_\mu] = s_\mu$, and the Schur functions form a basis in Sym (see Chapter 2).

Another corollary is the following fact: for an arbitrary constant c and any $k = 1, 2, \ldots$, the sequence of polynomials

$$p^*_{k,c;N}(x_1, \ldots, x_N) = \sum_{i=1}^{N} \left((x_i - i + c)^k - (-i + c)^k\right), \qquad N = 1, 2, \ldots,$$

determines an element $p^*_{k,c} \in \mathrm{Sym}^{*k}$. Note also that $[p^*_{k,c}] = p_k$. We consider the elements $p^*_{k,c}$ as shifted analogs of the power sums $p_k \in \mathrm{Sym}$. Proposition 2.9 implies that the elements $p^*_{1,c}, p^*_{2,c}, \ldots$ are algebraically independent generators of the algebra Sym^*, so that $\mathrm{Sym}^* = \mathbb{R}[p^*_{1,c}, p^*_{2,c}, \ldots]$.

Note that if $(x_1, x_2, \ldots)$ is an infinite vector with finitely many nonzero coordinates then we may write

$$p^*_{k,c}(x_1, x_2, \ldots) = \sum_{i=1}^{\infty} \left((x_i - i + c)^k - (-i + c)^k\right),$$

because the sum is actually finite. More generally, we may define the value at $(x_1, x_2, \ldots)$ for any element of Sym^*.

6.3 Modified Frobenius Coordinates

Recall that in Chapter 1 we introduced the Frobenius notation $\lambda = (p_1, \ldots, p_d \mid q_1, \ldots, q_d)$ for a Young diagram λ. Now it will be convenient for us to introduce the *modified Frobenius coordinates*, which differ from the conventional ones by adding $\frac{1}{2}$:

$$a_i = p_i + \tfrac{1}{2} = \lambda_i - i + \tfrac{1}{2}, \quad b_i = q_i + \tfrac{1}{2} = \lambda'_i - i + \tfrac{1}{2}, \qquad i = 1, \ldots, d,$$

[1] This notation is a little bit ambiguous because the highest term $[f^*]$ also depends on the choice of index k such that f^* is contained in the kth term of the filtration. However, the choice of k will always be clear from the context.

where, as before, d denotes the number of diagonal boxes in λ. Thus, the modified Frobenius coordinates are positive half-integers. Note that their sum equals $|\lambda|$, the number of boxes in λ. From now on we will change the Frobenius notation and write $\lambda = (a_1, \dots, a_d \mid b_1, \dots, b_d)$.

Proposition 6.6 *Let u be a formal variable. For any Young diagram $\lambda = (a_1, \dots, a_d \mid b_1, \dots, b_d)$ the following identity holds:*

$$\prod_{i=1}^{\ell} \frac{u + i - \frac{1}{2}}{u - \lambda_i + i - \frac{1}{2}} = \prod_{i=1}^{d} \frac{u + b_i}{u - a_i} \quad \text{for any } \ell \geq \ell(\lambda). \tag{6.4}$$

Proof Let $\square$ range over the boxes of λ and let $c(\square)$ denote the *content* of $\square$, i.e., $c(\square) = j - i$, where i and j are the row and column coordinates of $\square$. We have

$$\prod_{i=1}^{\ell} \frac{u + i - \frac{1}{2}}{u - \lambda_i + i - \frac{1}{2}} = \prod_{i=1}^{l} \frac{u + i - \frac{1}{2}}{u + i - \frac{3}{2}} \, \frac{u + i - \frac{3}{2}}{u + i - \frac{5}{2}} \cdots \frac{u + i - \lambda_i + \frac{1}{2}}{u + i - \lambda_i - \frac{1}{2}}$$

$$= \prod_{\square \in \lambda} \frac{u - c(\square) + \frac{1}{2}}{u - c(\square) - \frac{1}{2}} = \prod_{i=1}^{d} \prod_{\square \in \text{hook}_i} \frac{u - c(\square) + \frac{1}{2}}{u - c(\square) - \frac{1}{2}},$$

where hook_i stands for the ith diagonal hook in λ. As $\square$ ranges over hook_i, the quantity $c(\square) - \frac{1}{2}$ ranges from $-b_i$ to $a_i - 1$. From this we conclude that the product over hook_i equals $(u + b_i)/(u - a_i)$, which completes the proof. $\square$

Let us set

$$p_k^*(x_1, x_2, \dots) = p_{k,\frac{1}{2}}^*(x_1, x_2, \dots) = \sum_{i=1}^{\infty} \left((x_i - i + \tfrac{1}{2})^k - (-i + \tfrac{1}{2})^k \right), \tag{6.5}$$

where $k = 1, 2, \dots$.

Proposition 6.7 *For any Young diagram $\lambda = (a_1, \dots, a_d \mid b_1, \dots, b_d)$*

$$p_k^*(\lambda_1, \lambda_2, \dots) = \sum_{i=1}^{d} \left(a_i^k - (-b_i)^k \right), \qquad k = 1, 2, \dots.$$

Proof Write identity (6.4) as $L(u) = R(u)$. Here $L(u)$ and $R(u)$ are rational functions in variable u, which are regular at the point $u = \infty$ and take the value 1 at this point. Therefore, their logarithms are well defined about $u = \infty$. Expand both sides of the identity $\log L(u) = \log R(u)$ into a Taylor series with respect to variable u^{-1}. On one hand,

$$\log L(u) = \sum_{k=1}^{\infty} \frac{p_k^*(\lambda_1, \lambda_2, \dots)}{k} u^{-k},$$

and on the other hand,

$$\log R(u) = \sum_{k=1}^{\infty} \frac{\sum_{i=1}^{d} \left(a_i^k - (-b_i)^k\right)}{k} u^{-k}.$$

Equating the corresponding coefficients gives us the desired equality. □

6.4 The Embedding $\mathbb{Y}_n \to \Omega$ and Asymptotic Bounds

Definition 6.8 Given a Young diagram $\lambda = (a_1, \dots, a_d \mid b_1, \dots, b_d) \in \mathbb{Y}_n$, $n \geq 1$, set

$$\tfrac{1}{n}\omega_\lambda := \left(\frac{a_1}{n}, \dots, \frac{a_d}{n}, 0, 0, \dots; \frac{b_1}{n}, \dots, \frac{b_d}{n}, 0, 0, \dots\right). \tag{6.6}$$

Note that this is an element of Ω. Therefore, we get, for every $n = 1, 2, \dots$, a map $\mathbb{Y}_n \to \Omega$, which is evidently injective.

Note that $\frac{1}{n}\omega_\lambda$ lies in the dense subset $\Omega_0 \subset \Omega$ and, as $n \to \infty$, the image of the finite set $\mathbb{Y}_n$ becomes more and more dense in Ω (see Exercise 6.1). The embeddings $\mathbb{Y}_n \to \Omega$ will play the role similar to that of the maps $(k, l) \mapsto k/(k + l) \in [0, 1]$ in the context of the Pascal graph.

Recall that in Proposition 3.9 we defined an algebra morphism turning every element $f \in \mathrm{Sym}$ into a continuous function $f^\circ(\omega)$ on Ω.

Proposition 6.9 *Let $f^* \in \mathrm{Sym}^{*m}$ be an arbitrary shifted symmetric function of degree $\leq m$, $f = [f^*] \in \mathrm{Sym}^m$ be its highest term, and f° be the corresponding function on Ω. For any $\lambda \in \mathbb{Y}_n$, we have*

$$\frac{1}{n^m} f^*(\lambda_1, \lambda_2, \dots) = f^\circ(\tfrac{1}{n}\omega_\lambda) + O(\tfrac{1}{n}), \qquad n := |\lambda|, \tag{6.7}$$

where the $O(\frac{1}{n})$ bound for the remainder depends only on f^* and it is uniform in $\lambda \in \mathbb{Y}$.

Proof Assume first $f^* = p_k^*$ with $k = 1, 2, 3, \dots$. Then $f = p_k$, and Proposition 6.7 says that

$$\frac{1}{n^k} p_k^*(\lambda_1, \lambda_2, \dots) = \frac{1}{n^k} \sum_{i=1}^{d} \left(a_i^k - (-b_i)^k\right).$$

On the other hand, comparing (3.7) and (6.6) we see that

$$p_k^\circ(\tfrac{1}{n}\omega_\lambda) = \frac{1}{n^k} \sum_{i=1}^{d} \left(a_i^k - (-b_i)^k\right), \qquad k = 2, 3, \dots.$$

The same also holds for $k = 1$, because $\sum(a_i + b_i) = n$. Therefore,

$$\frac{1}{n^k} p_k^*(\lambda_1, \lambda_2, \dots) = p_k^\circ(\tfrac{1}{n}\omega_\lambda), \qquad k = 1, 2, \dots .$$

It follows that if f^* is a monomial in generators $p_1^*, p_2^*, \dots$ and $\deg f^* = m$, then we again get an exact equality:

$$\frac{1}{n^m} f^*(\lambda_1, \lambda_2, \dots) = f^\circ(\tfrac{1}{n}\omega_\lambda).$$

Finally, an arbitrary element $f^* \in \mathrm{Sym}^*$ of degree $\le m$ can be written as a linear combination of monomials $f_1^*, f_2^*, \dots$ in the generators:

$$f^* = c_1 f_1^* + c_2 f_2^* + \cdots ,$$

where $c_1, c_2, \dots$ are some coefficients. Hence,

$$\frac{1}{n^m} f^*(\lambda_1, \lambda_2, \dots) = \sum_j c_j \frac{n^{\deg f_j^*}}{n^m} f_j^\circ(\tfrac{1}{n}\omega_\lambda)$$

$$= f^\circ(\tfrac{1}{n}\omega_\lambda) + \sum_{j:\, \deg f_j^* < m} c_j \frac{n^{\deg f_j^*}}{n^m} f_j^\circ(\tfrac{1}{n}\omega_\lambda).$$

The latter sum representing the remainder in (6.7) is $O(\frac{1}{n})$, because every function f_j° is bounded on Ω. □

Recall that $\dim(\mu, \lambda)$ denotes the number of paths in the Young graph going from μ to λ and increasing the level number by one on each step (Definition 6.1).

Theorem 6.10 *For any fixed $\mu \in \mathbb{Y}$ and varying $\lambda \in \mathbb{Y}$, we have*

$$\frac{\dim(\mu, \lambda)}{\dim \lambda} = s_\mu^\circ(\tfrac{1}{n}\omega_\lambda) + O(\tfrac{1}{n}), \qquad n := |\lambda|, \tag{6.8}$$

where the bound of the remainder depends only on μ and is uniform in $\lambda \in \mathbb{Y}$.

Proof Indeed, Proposition 6.5 says that

$$\frac{\dim(\mu, \lambda)}{\dim \lambda} = \frac{s_\mu^*(\lambda_1, \lambda_2, \dots)}{n^{\downarrow m}}$$

and Proposition 6.9 applied to $f^* = s_\mu^*$ gives

$$\frac{s_\mu^*(\lambda_1, \lambda_2, \dots)}{n^m} = s_\mu^\circ(\tfrac{1}{n}\omega_\lambda) + O(\tfrac{1}{n}).$$

Therefore,

$$\frac{\dim(\mu, \lambda)}{\dim \lambda} = \frac{n^m}{n^{\downarrow m}} \left(s_\mu^\circ(\tfrac{1}{n}\omega_\lambda) + O(\tfrac{1}{n})\right) = s_\mu^\circ(\tfrac{1}{n}\omega_\lambda) + O(\tfrac{1}{n}),$$

because

$$\frac{n^m}{n^{\downarrow m}} = 1 + O(\tfrac{1}{n})$$

and the function s°_μ is bounded on Ω. □

Let us sum up the mechanism of the proof of Theorem 6.10: we fix μ and consider the ratio $\dim(\mu,\lambda)/\dim\lambda$ as a function in λ. It turns out that after an appropriate normalization (multiplication by $n^{\downarrow m}$) we get a sufficiently "regular" function on $\mathbb{Y}$ (we mean the function $s^*_\mu(\lambda) = s^*_\mu(\lambda_1, \lambda_2, \ldots)$). Moreover, the linear span $\mathbb{A}$ of all such functions indexed by μs turns out to be an algebra (it is isomorphic to algebra Sym*), and we are able to understand its structure rather well. This finally enables us to get the required estimate for the asymptotics of our functions as λ goes to infinity.

Let us give a formal definition of $\mathbb{A}$:

Definition 6.11 Set

$$H(u;\lambda) = \prod_{i=1}^{\infty} \frac{u + i - \frac{1}{2}}{u - \lambda_i + i - \frac{1}{2}}, \qquad u \in \mathbb{C}, \quad \lambda \in \mathbb{Y}.$$

The product is actually finite, because $\lambda_i = 0$ provided that $i > \ell(\lambda)$, and it is a rational function in $u \in \mathbb{C}$ taking value 1 at $u = \infty$. Consider the Taylor expansion of $H(u;\lambda)$ at the point $u = \infty$ with respect to variable u^{-1}. The algebra $\mathbb{A}$ of *regular functions* on $\mathbb{Y}$ is defined as the unital algebra over $\mathbb{R}$ generated by the coefficients of that expansion.

Obviously, instead of $H(u;\lambda)$, we can equally well deal with the Taylor expansion at $u = \infty$ of $\log H(u;\lambda)$. Introduce the functions $p_1(\lambda), p_2(\lambda), \ldots$ on $\mathbb{Y}$ from the expansion

$$\log H(u;\lambda) = \sum_{k=1}^{\infty} \frac{p_k(\lambda)}{k} u^{-k}.$$

Then, by virtue of Propositions 6.6 and 6.7,

$$p_k(\lambda) = \sum_{i=1}^{\infty} [(\lambda_i - i + \tfrac{1}{2})^k - (-i + \tfrac{1}{2})^k] = \sum_{j=1}^{d} \left(a_j^k - (-b_j)^k\right).$$

This means that regular functions on $\mathbb{Y}$ are both shifted symmetric functions in the row coordinates $\lambda_1, \lambda_2, \ldots$ of a Young diagram $\lambda \in \mathbb{Y}$ and supersymmetric functions (see Exercise 2.17, above) in its modified Frobenius coordinates.

Remark 6.12 By the very definition of Sym*, it is a filtered algebra such that the associated graded algebra gr Sym* is canonically isomorphic to Sym. As pointed out in [80, Remark 1.7], Sym* can be viewed as a deformation of Sym. On the other hand, there are natural isomorphisms

$$\mathrm{Sym}^* \to \mathbb{A} \leftarrow \mathrm{Sym}$$

which make it possible to choose a natural algebra isomorphism between Sym* and Sym, consistent with the identification gr Sym* = Sym.

Indeed, it is easily seen that the functions $p_1(\lambda), p_2(\lambda), \dots$ on $\mathbb{Y}$ are algebraically independent. This implies that the correspondence

$$\mathrm{Sym}^* \ni p_k^* \to p_k(\,\cdot\,) \in \mathbb{A}$$

is an algebra isomorphism $\mathrm{Sym}^* \to \mathbb{A}$. Next, the algebra of supersymmetric functions is simply an incarnation of the algebra of symmetric functions Sym (see Exercise 2.17), so that we get an isomorphism $\mathrm{Sym} \to \mathbb{A}$ turning the power-sum generators $p_k \in \mathrm{Sym}$ into the super-power-sums $p_k(\lambda)$.

For more detail see Olshanski, Regev, and Vershik [97], [98].

6.5 Integral Representation of Coherent Systems: Proof

Here we deduce Theorem 3.12 from Theorem 6.10. Recall the statement of Theorem 3.12: the relation

$$M^{(n)}(\lambda) = \dim\lambda \int_\Omega s_\lambda^\circ(\omega) P(d\omega), \qquad n = 0, 1, 2, \dots, \qquad \lambda \in \mathbb{Y}_n, \quad (6.9)$$

determines a bijective correspondence $\{M^{(n)}\} \leftrightarrow P$ between coherent systems of distributions on $\mathbb{Y}$ and probability measures on Ω.

Proof *Step 1.* We start with a generalization of Definition 3.3:

Definition 6.13 With every pair $n > m$ of nonnegative integers we associate the $\mathbb{Y}_n \times \mathbb{Y}_m$-matrix Λ^n_m with entries

$$\Lambda^n_m(\lambda, \mu) = \frac{\dim\mu \cdot \dim(\mu, \lambda)}{\dim\lambda}, \qquad \lambda \in \mathbb{Y}_n, \quad \mu \in \mathbb{Y}_m. \quad (6.10)$$

By the definition of the relative dimension (Definition 6.1), $\Lambda^n_m(\lambda, \mu)$ vanishes unless λ contains μ. Note also that when $m = n - 1$, the above definition agrees with the definition of the matrix Λ^n_{n-1} given above (see Definition 3.3).

More generally, for $n - m \geq 2$ the matrix Λ^n_m factorizes into a product of $n - m$ matrices of the form Λ^i_{i-1}, where i ranges from n to $m + 1$:

$$\Lambda^n_m = \Lambda^n_{n-1}\Lambda^{n-1}_{n-2} \cdots \Lambda^{m+1}_m. \quad (6.11)$$

Indeed, this easily follows from the interpretation of the relative dimension in terms of monotone paths in the Young graph.

The first consequence of (6.11) is that Λ^n_m is a stochastic matrix, for it is a product of stochastic matrices.

Next, applying (6.11) and iterating the coherence relation (3.2) we get the equality

$$M^{(n)}\Lambda^n_m = M^{(m)}, \qquad n > m, \tag{6.12}$$

which holds for any coherent system $\{M^{(n)}\}$.

Step 2. We are going to extend the definition of the stochastic matrix Λ^n_m to the case "$n = \infty$". The corresponding object, denoted as Λ^∞_m, is no longer a matrix, but it can be defined as a kernel of format $\Omega \times \mathbb{Y}_m$. We set

$$\Lambda^\infty_m(\omega, \mu) := \dim \mu \cdot s^\circ_\mu(\omega), \qquad \mu \in \mathbb{Y}_m. \tag{6.13}$$

Obviously, the kernel is continuous in the first variable, for $s^\circ_\mu(\omega)$ is a continuous function on Ω.

In the above notation, Theorem 6.10 can be restated as the asymptotic relation

$$\Lambda^n_m(\lambda, \mu) = \Lambda^\infty_m(\tfrac{1}{n}\omega_\lambda, \mu) + O(\tfrac{1}{n}), \qquad \lambda \in \mathbb{Y}_n. \tag{6.14}$$

If we agree to identify λ with $\frac{1}{n}\omega_\lambda$, then (6.14) shows that the kernel Λ^∞_m may be viewed as a large-n limit of the matrices Λ^n_m.

Let us develop this claim further. Every point $\omega \in \Omega$ can be approximated by points of the form $\frac{1}{n}\omega_{\lambda(n)}$ with $\lambda(n) \in \mathbb{Y}_n$, as $n \to \infty$ (Exercise 6.1). Combining this fact with continuity of the kernel we get from (6.14) the following conclusions:

First, Λ^∞_m is a stochastic kernel, meaning that for every $\omega \in \Omega$ and every $m = 0, 1, 2, \dots$,

$$\Lambda^\infty_m(\omega, \mu) \ge 0, \qquad \sum_{\mu \in \mathbb{Y}_m} \Lambda^\infty_m(\omega, \mu) = 1.$$

Second, for any $n > m$,

$$\Lambda^\infty_n \Lambda^n_m = \Lambda^\infty_m. \tag{6.15}$$

To see this, observe that, for $n' > n > m$,

$$\Lambda^{n'}_n \Lambda^n_m = \Lambda^{n'}_m \quad \text{(by virtue of (6.11))}$$

and then pass to the limit as $n' \to \infty$.

Step 3. Given a probability measure P on Ω, set

$$M^{(n)} := P\Lambda^\infty_n, \qquad n = 0, 1, 2, \dots .$$

The correspondence $P \mapsto \{M^{(n)}\}$ thus defined is just a reformulation of (6.9).

Since Λ_n^∞ is a stochastic kernel for every n, the $M^{(n)}$s are probability measures.

Next, the sequence $\{M^{(n)}\}$ is coherent. Indeed, (6.15) implies that $M^{(n)}\Lambda_m^n = M^{(m)}$ for any $n > m$.

Finally, the map $P \mapsto \{M^{(n)}\}$ is injective. Indeed, assume that $P\Lambda_n^\infty = P'\Lambda_n^\infty$ for all n. This means that

$$\int_\Omega s_\lambda^\circ(\omega)P(d\omega) = \int_\Omega s_\lambda^\circ(\omega)P'(d\omega)$$

for all λ. But the functions s_λ° span a dense subspace in $C(\Omega)$, whence $P = P'$.

Step 4. It remains to prove that the map $P \mapsto \{M^{(n)}\}$ is surjective. To do this, let us fix a coherent system $\{M^{(n)}\}$ and show that it comes from a probability measure P.

Let $P^{(n)}$ be the pushforward of $M^{(n)}$ under the embedding $\mathbb{Y}_n \to \Omega$. This is a probability measure on Ω. In more detail,

$$P^{(n)} = \sum_{\lambda \in \mathbb{Y}_n} M^{(n)}(\lambda)\langle \tfrac{1}{n}\omega_\lambda \rangle, \tag{6.16}$$

where $\langle \omega \rangle$ stands for the delta-measure at a point $\omega \in \Omega$.

Since the space Ω is compact and metrizable, any sequence of probability measures has partial weak limits, which also are probability measures. Let P be any such partial limit. We claim that

$$P\Lambda_m^\infty = M^{(m)}$$

for every m. Indeed, fix m and write the coherence relation in the form

$$(M^{(n)}\Lambda_m^n)(\mu) = M^{(m)}(\mu), \qquad n > m, \quad \mu \in \mathbb{Y}_m.$$

By virtue of (6.14) this implies

$$(P^{(n)}\Lambda_m^\infty)(\mu) = M^{(m)}(\mu) + O(\tfrac{1}{n}).$$

Passing to a limit as n goes to infinity along a suitable subsequence of indices n, we get the desired relation $P\Lambda_m^\infty = M^{(m)}$.

This completes the proof. □

The following result is a direct consequence of the above argument:

Theorem 6.14 (Approximation theorem) *Let P be a probability measure on Ω and $\{M^{(n)}\}$ be the corresponding coherent system. As $n \to \infty$, the measures $M^{(n)}$ approximate P in the sense that their pushforwards $P^{(n)}$ determined by* (6.16) *weakly converge to P.*

6.6 The Vershik–Kerov Theorem

Definition 6.15 Assume we are given a sequence $\{\lambda(n) \in \mathbb{Y}_n\}$ of Young diagrams, $n = 1, 2, \dots$.

(i) Let us say that the corresponding normalized irreducible characters *converge* if, for any fixed $g \in S(\infty)$, there exists a limit

$$\lim_{n\to\infty} \frac{\chi^{\lambda(n)}(g)}{\dim \lambda(n)} = \chi(g). \tag{6.17}$$

(Note that the quantity $\chi^{\lambda(n)}(g)$ makes sense for n large enough: so large that g is contained in $S(n)$.)

(ii) Let us write $\lambda(n) \to \omega \in \Omega$ (in words, *the sequence* $\{\lambda(n)\}$ *converges to a point* $\omega \in \Omega$) if the points $\frac{1}{n}\omega_{\lambda(n)}$ converge to ω in the topology of Ω.

By the very definition of the map $\lambda \mapsto \frac{1}{n}\omega_\lambda$ (Definition 6.8), the convergence $\lambda(n) \to \omega$ means that the normalized modified Frobenius coordinates of $\lambda(n)$ completed by 0s converge to the respective coordinates α_i, β_i of ω. Equivalently,

$$\lim_{n\to\infty} \frac{\lambda(n)_i}{n} = \alpha_i, \quad \lim_{n\to\infty} \frac{(\lambda(n))'_i}{n} = \beta_i, \qquad i = 1, 2, \dots . \tag{6.18}$$

We are going to prove the following result.

Theorem 6.16 (Vershik–Kerov theorem) *Normalized irreducible characters indexed by a sequence* $\{\lambda(n) \in \mathbb{Y}_n\}$ *converge if and only if the sequence* $\{\lambda(n)\}$ *converges to a point* $\omega \in \Omega$. *Then the limit function* χ *coincides with the extreme character* χ^ω *indexed by* ω.

We argued in the Introduction that the irreducible characters χ^λ of finite symmetric groups are "special functions", while the characters χ^ω of $S(\infty)$ are "elementary" ones. Theorem 6.16 makes this fact less surprising, for it often happens that asymptotics of special functions are described by elementary functions.

Proof By the Young branching rule (Proposition 1.4, item (i)), the restriction of the irreducible character $\chi^{\lambda(n)}$ to the subgroup $S(m) \subset S(n)$, where $m < n$, can be written as

$$\operatorname{Res}^{S(n)}_{S(m)} \chi^{\lambda(n)} = \sum_{\mu\in\mathbb{Y}_m} \dim(\mu, \lambda(n))\chi^\mu.$$

Equivalently,

$$\operatorname{Res}^{S(n)}_{S(m)} \frac{\chi^{\lambda(n)}}{\dim \lambda(n)} = \sum_{\mu\in\mathbb{Y}_m} \frac{\dim(\mu, \lambda(n))}{\dim \lambda(n)} \chi^\mu.$$

It follows that the existence of a limit (6.17) is equivalent to the existence of the limits

$$\lim_{n\to\infty} \frac{\dim(\mu, \lambda(n))}{\dim \lambda(n)}, \qquad \forall \mu \in \mathbb{Y}. \tag{6.19}$$

Therefore, it suffices to show that the latter condition is equivalent to convergence of the sequence $\{\frac{1}{n}\omega_{\lambda(n)}\}$ in Ω.

Assume that $\frac{1}{n}\omega_{\lambda(n)} \to \omega \in \Omega$ as $n \to \infty$. By Theorem 6.10,

$$\frac{\dim(\mu, \lambda(n))}{\dim \lambda(n)} = s^\circ_\mu(\tfrac{1}{n}\omega_{\lambda(n)}) + O(\tfrac{1}{n}).$$

Since the function $s^\circ_\mu(\omega)$ is continuous, the limit in (6.19) exists and equals $s^\circ_\mu(\omega)$.

Conversely, assume that the limits in (6.19) exist. By compactness of Ω, we can choose a subsequence of $\{\lambda(n)\}$ such that the corresponding points in Ω converge to some point $\omega \in \Omega$. Then, by the first part of the proof, (6.19) holds along this subsequence.

Since the linear span of the functions $s^\circ_\mu(\omega)$ separates points of Ω (see claim (ii) of Proposition 3.9), all limit points of the sequence $\{\frac{1}{n}\omega_{\lambda(n)}\}$ must coincide. □

6.7 Exercises

Exercise 6.1 Show that for any point $\omega \in \Omega$ there exists a sequence $\{\lambda(n) \in \mathbb{Y}_n\}$ of Young diagrams such that $\frac{1}{n}\omega_{\lambda(n)} \to \omega$ as $n \to \infty$.

Exercise 6.2 To any point $\omega \in \Omega$ one assigns an atomic probability measure ν_ω on $[-1, 1]$, called the *Thoma measure* associated to ω, by setting

$$\nu_\omega = \sum_{i=1}^{\infty} \alpha_i \delta_{\alpha_i} + \sum_{i=1}^{\infty} \beta_i \delta_{-\beta_i} + \gamma \delta_0 ,$$

where, as usual, $\alpha = (\alpha_1, \alpha_2, \dots)$ and $\beta = (\beta_1, \beta_2, \dots)$ are the Thoma parameters corresponding to ω, γ is given by (3.9), and δ_x stands for the Dirac delta-measure at a point $x \in \mathbb{R}$.

(a) Let

$$q_k = q_k(\omega) = \int x^k \nu_\omega(dx), \qquad k = 0, 1, 2, \dots,$$

denote the moments of the Thoma measure. Check the equality

$$p^\circ_i(\omega) = q_{i-1}(\omega), \qquad i = 1, 2, \dots .$$

Since $q_0 \equiv 1$, this provides one more justification of the agreement $p^\circ_1 \equiv 1$.

(b) Show that the correspondence $\omega \mapsto \nu_\omega$ gives a homeomorphism of Ω onto a closed subspace of $\mathscr{M}_1[-1, 1]$, the space of probability measures on $[-1, 1]$ equipped with the weak topology.

Thus, we obtain a nice interpretation of the subalgebra $\mathrm{Sym}^\circ \subset C(\Omega)$ and of its generators $p_i^\circ(\omega)$: this is simply the algebra generated by the moments $q_0 = 1, q_1, q_2, \dots$.

Exercise 6.3 (a) Show that the algebra $\mathbb{A}$ of regular functions on $\mathbb{Y}$ is stable under the change of variable $\lambda \mapsto \lambda'$ (transposition of diagram λ). Deduce from this that elements of $\mathbb{A}$ are shifted symmetric functions in the "dual variables", the column lengths λ_i'.

(b) On the contrary, the algebra of conventional symmetric functions in variables $\lambda_1, \lambda_2, \dots$ does not possess this property; it is not consistent with transposition $\lambda \to \lambda'$.

Exercise 6.4 Besides $H(u; \lambda)$ (see Definition 6.11), there is one more "reasonable" function in variables u and λ with obvious shifted symmetry property in λ. It is given by the infinite series

$$\sum_{i=1}^{\infty} e^{(\lambda_i - i + \frac{1}{2})u}.$$

(a) Show that the series converges in the right half-plane $\Re u > 0$.

(b) Show that

$$\sum_{i=1}^{\infty} e^{(\lambda_i - i + \frac{1}{2})u} = \frac{1}{u} + \sum_{k=1}^{\infty} \widehat{p}_k(\lambda) \frac{u^k}{k!}, \qquad \lambda \in \mathbb{Y}, \quad \Re u > 0,$$

where

$$\widehat{p}_k(\lambda) = p_k(\lambda) + (1 - 2^{-k})\zeta(-k)$$

and $\zeta(\cdot)$ is Riemann's zeta-function.

Exercise 6.5 Define the *content power sums* as the following functions on $\mathbb{Y}$:

$$\pi_k(\lambda) := \sum_{\square \in \lambda} (c(\square))^k, \qquad k = 0, 1, 2, \dots,$$

in particular,

$$\pi_0(\lambda) := |\lambda|.$$

Notice that, unlike the ordinary power sums, these functions are enumerated from $k = 0$.

(a) Show that the content power sums belong to the algebra $\mathbb{A}$ and form a system of its algebraically independent generators. Thus, $\mathbb{A}$ can be identified with $\mathbb{R}[\pi_0, \pi_1, \pi_2, \dots]$.

(b) Show that π_k has degree $k+1$. More precisely,

$$\pi_k = \frac{p_{k+1}}{k+1} + \text{lower degree terms}, \qquad k = 0, 1, 2, \dots .$$

Exercise 6.6 The identification of partitions $\lambda = (\lambda_1, \lambda_2, \dots)$ with Young diagrams makes it possible to introduce various encodings of λs. First of all, one may pass from the row lengths λ_i to the column lengths λ'_i. Next, one can consider the classical or modified Frobenius coordinates, which are consistent with the symmetry $\lambda \to \lambda'$. In a number of cases they turn out to be more effective than the row or column lengths. Here we discuss one more useful encoding of Young diagrams, invented by Kerov [62].

(a) Recall that we draw Young diagrams in a quarter plane according to the so-called "English picture" (Macdonald [72]), which means that the first coordinate axis is directed downwards and the second axis is directed to the right. Denote the coordinates along these axes as r and s, respectively. The boundary of a diagram λ is a broken line, which we imagine as a directed path coming from $+\infty$ along the s-axis, next turning several times alternatingly down and to the left, and finally going away to $+\infty$ along the r-axis. The turning points, called the *corners* of λ, are of two types: the *inner corners*, where the path switches from the horizontal direction to the vertical one, and the *outer corners*, where the direction is switched from vertical to horizontal. Observe that the corners of both types interlace and that the number of inner corners always exceeds by 1 that of the outer corners. Let $2k-1$ be the total number of the corners, (r_i, s_i), $1 \le i \le k$, be the coordinates of the inner corners, and (r'_j, s'_j), $1 \le j \le k-1$, be the coordinates of the outer corners. Set

$$x_i = s_i - r_i, \quad 1 \le i \le k, \qquad y_i = s'_j - r'_j, \quad 1 \le j \le k-1.$$

By the very construction these are interlacing integers:

$$x_1 > y_1 > x_2 > \cdots > \cdots > x_{k-1} > y_{k-1} > x_k. \tag{6.20}$$

We call these numbers the *Kerov interlacing coordinates* of λ. Obviously, these coordinates determine the initial diagram λ uniquely.

(b) We will use the notation $\lambda = (X; Y)$, where $X = \{x_i\}$ and $Y = \{y_j\}$. Observe that there is a natural bijective correspondence between the inner corners of λ (and thus the points $x_i \in X$) and the boxes that can be appended to λ. Likewise, the outer corners (and thus the points $y_j \in Y$) correspond to the boxes that can be removed from λ.

Describe the transformation of $(X; Y)$ when a box is appended or removed. A useful device to prove various statements involving the Kerov interlacing coordinates of λ is to argue by induction on $|\lambda|$ using these transformations.

(c) Show that two interlacing sequences of integers (6.20) come from a Young diagram if and only if

$$\sum_{i=1}^{k} x_i - \sum_{j=1}^{k-1} y_j = 0.$$

Hint: use the device mentioned in item (b) above.

(d) Prove the formula

$$|\lambda| = \sum_{1 \le i \le j \le k-1} (x_i - y_i)(y_j - x_{j+1}).$$

Exercise 6.7 Here we aim to demonstrate an alternative approach to the algebra $\mathbb{A}$ using Kerov's interlacing coordinates $(X; Y)$ of a diagram $\lambda \in \mathbb{Y}$ introduced above.

(a) Let $\widetilde{\mathrm{Sym}}$ stand for a copy of the algebra of symmetric functions. Denote by $\widetilde{p}_n$ and $\widetilde{h}_n$ the power sums and the complete homogeneous functions viewed as elements of $\widetilde{\mathrm{Sym}}$. We are going to define an algebra morphism of $\widetilde{\mathrm{Sym}}$ into the algebra of functions on $\mathbb{Y}$. To do this it suffices to specify the functions $\widetilde{h}_n(\lambda)$ that will serve as the images of the generators $\widetilde{h}_n \in \widetilde{\mathrm{Sym}}$. We define them by setting

$$1 + \sum_{n=1}^{\infty} \widetilde{h}_n(\lambda) u^{-n} = \text{the expansion of } \widetilde{H}(u; \lambda) \text{ about } u = \infty, \qquad \lambda \in \mathbb{Y},$$

where

$$\widetilde{H}(u; \lambda) = \frac{u \prod_{j=1}^{k-1} (u - y_j)}{\prod_{i=1}^{k} (u - x_i)}, \qquad \lambda \in \mathbb{Y}.$$

Observe that $\widetilde{H}(u; \lambda)$ is a rational function in u such that $\widetilde{H}(u; \lambda) = 1 + O(u^{-1})$ about $u = \infty$, so that the definition is correct.

Show that the functions $\widetilde{p}_n(\lambda)$ corresponding to the generators $\widetilde{p}_n \in \widetilde{\mathrm{Sym}}$ are given by

$$\widetilde{p}_n(\lambda) = \sum_{i=1}^{k} x_i^n - \sum_{j=1}^{k-1} y_j^n, \qquad n = 1, 2, \ldots, \qquad \lambda \in \mathbb{Y}.$$

It follows that the function $\widetilde{f}(\lambda)$ on $\mathbb{Y}$ corresponding to an arbitrary element $\widetilde{f} \in \widetilde{\mathrm{Sym}}$ is obtained by evaluating the supersymmetric function $\widetilde{f}(\cdot, \cdot)$ at $(X; -Y)$. Here we mean the definition of supersymmetric functions given in Exercise 2.17 above.

Observe that

$$\widetilde{p}_1(\lambda) \equiv 0,$$

by virtue of Exercise 6.6(c).

(b) Prove the formula

$$\widetilde{H}(u; \lambda) = \frac{H(u - \frac{1}{2}; \lambda)}{H(u + \frac{1}{2}; \lambda)}, \qquad \lambda \in \mathbb{Y},$$

relating the definition of item (a) above to Definition 6.11.

(c) Deduce from (b) that the functions $\widetilde{p}_2(\lambda), \widetilde{p}_3(\lambda), \ldots$ belong to the algebra $\mathbb{A}$ and are related to the functions $p_1(\lambda), p_2(\lambda)$ by $(n = 2, 3, \ldots)$

$$\widetilde{p}_n = \sum_{j=0}^{[\frac{n-1}{2}]} \binom{n}{2j+1} 2^{-2j}\, p_{n-1-2j} = np_{n-1} + \text{lower degree terms}$$

(recall that $\widetilde{p}_1 \equiv 0$).

It follows that the algebra $\mathbb{A}$ coincides with image of algebra $\widetilde{\mathrm{Sym}}$ under the morphism defined in (a). The kernel of this morphism is the principal ideal generated by $\widetilde{p}_1 \in \widetilde{\mathrm{Sym}}$, so that the $\mathbb{A}$ can be viewed as the quotient $\widetilde{\mathrm{Sym}}/\widetilde{p}_1\widetilde{\mathrm{Sym}}$. For a generalization, see Olshanski [94].

Exercise 6.8 Summing up, the functions $f(\lambda)$ on $\mathbb{Y}$ belonging to the algebra $\mathbb{A}$ can be described in four different ways:

(1) as shifted symmetric functions in $\lambda_1, \lambda_2, \ldots$;

(2) as supersymmetric functions in $(a; b)$, the modified Frobenius coordinates of λ;

(3) as symmetric functions in the box contents of λ (one should also add as an additional generator the function $\pi_0(\lambda) = |\lambda|$);

(4) as supersymmetric functions in $(X; -Y)$, where $X = (x_1, \ldots, x_k)$ and $Y = (y_1, \ldots, y_{k-1})$ are Kerov's interlacing coordinates of λ.

Exercise 6.9 Consider the algebra isomorphism $\mathrm{Sym}^* \to \mathrm{Sym}$ defined in Remark 6.12. The image under this isomorphism of the shifted Schur function s^*_μ is called the *Frobenius–Schur function* and denoted as FS_μ.

Below we will freely pass from Sym to $\mathbb{A}$ and *vice versa*, using the isomorphism between these two algebras. For any $f \in \mathrm{Sym}$ we will denote by $f(\lambda)$ the corresponding function from $\mathbb{A}$.

(a) The Frobenius–Schur functions can be characterized by the property

$$FS_\mu(\lambda) = n^{\downarrow m} \frac{\dim(\mu, \lambda)}{\dim \lambda}, \qquad \lambda \in \mathbb{Y}, \quad n = |\lambda|, \quad m = |\mu|.$$

(b) Show that the highest homogeneous component of FS_μ is the Schur symmetric function s_μ.

(c) Prove the symmetry relation $FS_\mu(\lambda') = FS_{\mu'}(\lambda)$.

Exercise 6.10 Recall the notation χ^λ_ρ for the symmetric group characters, see Chapter 2.

(a) Show that for any partition ρ there exists a unique element $p^\#_\rho \in \mathrm{Sym} = \mathbb{A}$ such that for any $\lambda \in \mathbb{Y}$ with $n := |\lambda| \geq |\rho| =: m$

$$n^{\downarrow m} \frac{\chi^\lambda_{\rho \cup 1^{n-m}}}{\dim \lambda} = p^\#_\rho(\lambda).$$

Moreover,

$$p^\#_\rho = p_\rho + \text{lower degree terms}.$$

(b) Show that the elements $p^\#_\rho$ are related to the Frobenius–Schur functions FS_λ exactly as the elements p_ρ are related to the conventional Schur functions s_λ:

$$p^\#_\rho = \sum_{\lambda:\, |\lambda| = |\rho|} \chi^\lambda_\rho FS_\lambda.$$

(c) Prove the asymptotic relation

$$\frac{\chi^\lambda_{\rho \cup 1^{|\lambda| - |\rho|}}}{\dim \lambda} = p^\circ_\rho(\tfrac{1}{n}\omega_\lambda) + O(|\lambda|^{-\frac{1}{2}}).$$

6.8 Notes

The material related to the shifted Schur functions and the algebra Sym* of shifted symmetric functions is taken from Okounkov–Olshanski [80] (for further development of the subject see Olshanski–Regev–Vershik [97], [98]).

Shifted symmetric functions in the row coordinates $\lambda_1, \lambda_2, \ldots$ of Young diagrams $\lambda \in \mathbb{Y}$ appeared in Olshanski [83] and [88], while supersymmetic functions in the modified Frobenius coordinates of λ originated in Vershik–Kerov [122]. The remarkable fact that both kinds of functions are actually the same was observed in the note by Kerov–Olshanski [65]. That note also contained the definition of the algebra $\mathbb{A}$.

Note that in [65], shifted symmetric functions were called quasi-symmetric functions; unfortunately, at that time the authors did not know that this term had been introduced earlier by Gessel and had a different meaning.

In the classical representation theory of finite groups, irreducible characters are usually regarded as functions on a group or, which is essentially equivalent, on the set of its conjugacy classes. In the asymptotic approach of Vershik–Kerov [122], the picture is reversed, and the normalized irreducible characters are regarded as functions on the set $\mathbb{Y}$ of Young diagrams λ (the representation labels). This approach is at the heart of our proof of Thoma's theorem; only we prefer to deal with the relative dimension instead of the normalized character. The two quantities are closely related; see Okounkov–Olshanski [80].

Quite different proofs of Thoma's theorem were given in Okounkov's dissertation (see [78] and [79]) and in a recent paper [26] by Bufetov and Gorin.

The idea of reverting the character table of symmetric groups was pushed further by Kerov and led to a series of works devoted to the study of the so-called Kerov character polynomials; see Féray [40] and references therein.

7

Boundaries and Gibbs Measures on Paths

Thoma's theorem and de Finetti's theorem can be placed in the context of a much more general formalism of finding the boundary of a graded graph. Let us briefly describe the general setting.

7.1 The Category $\mathscr{B}$

About the notions used in this section, see Parthasarathy [100] and Meyer [74]. A *measurable space* (also called *Borel space*) is a set with a distinguished sigma-algebra of subsets. Denote by $\mathscr{B}$ the category whose objects are *standard* measurable spaces[1] and whose morphisms are Markov kernels. A morphism between two objects will be denoted by a dashed arrow, $X \dashrightarrow Y$, in order to emphasize that it is not an ordinary map. Recall that a (stochastic) *Markov kernel* $\Lambda : X \dashrightarrow Y$ between two measurable spaces X and Y is a function $\Lambda(a, A)$, where a ranges over X and A ranges over measurable subsets of Y, such that $\Lambda(a, \cdot)$ is a probability measure on Y for any fixed a and $\Lambda(\cdot, A)$ is a measurable function on X for any fixed A.

Below we use the short term *link* as a synonym of "Markov kernel". The composition of two links will be read from left to right: given $\Lambda : X \dashrightarrow Y$ and $\Lambda' : Y \dashrightarrow Z$, their composition $\Lambda\Lambda' : X \dashrightarrow Z$ is defined as

$$(\Lambda\Lambda')(x, dz) = \int_Y \Lambda(x, dy)\Lambda'(y, dz),$$

where $\Lambda(x, dy)$ and $\Lambda'(y, dz)$ symbolize the measures $\Lambda(x, \cdot)$ and $\Lambda'(y, \cdot)$, respectively.

[1] Standard measurable spaces are, up to isomorphism, of two types: (1) the finite set $\{1, \dots, n\}$ or the countably infinite set $\{1, 2, \dots\}$ with the sigma-algebra formed by arbitrary subsets; (2) the interval $[0, 1]$ with the sigma-algebra of Borel subsets. For more detail, see Parthasarathy [100, Chapter V].

A *projective system* in $\mathscr{B}$ is a family $\{V_i, \Lambda_i^j\}$ consisting of objects V_i indexed by elements of a linearly ordered set I (not necessarily discrete), together with links $\Lambda_i^j : V_j \dashrightarrow V_i$ defined for any couple $i < j$ of indices, such that for any triple $i < j < k$ of indices, one has $\Lambda_j^k \Lambda_i^j = \Lambda_i^k$.

A *limit object* of a projective system is understood in the categorical sense: this is an object $X = \varprojlim V_i$ together with links $\Lambda_i^\infty : X \dashrightarrow V_i$ defined for all $i \in I$, such that:

- $\Lambda_j^\infty \Lambda_i^j = \Lambda_i^\infty$ for all $i < j$;
- if an object Y and links $\widetilde{\Lambda}_i^\infty : Y \dashrightarrow V_i$ satisfy the similar condition, then there exists a unique link $\Lambda_X^Y : Y \dashrightarrow X$ such that $\widetilde{\Lambda}_i^\infty = \Lambda_X^Y \Lambda_i^\infty$.

General results concerning the existence and uniqueness of limit objects in $\mathscr{B}$ can be found in Winkler [134, Chapter 4]. See also Dynkin [34], [35], Kerov and Orevkova [68]. When the index set I is a subset of $\mathbb{R}$ and all spaces V_i are copies of one and the same space X, our definition of projective system turns into the classical notion of *transition function* on X (within inversion of order on I).

For a measurable space X we denote by $\mathscr{M}(X)$ the set of probability measures on X. It is itself a measurable space – the corresponding sigma-algebra is generated by sets of the form $\{\mu \in \mathscr{M}(X) : \mu(A) \in B\}$, where $A \subseteq X$ is a measurable and $B \subseteq \mathbb{R}$ is Borel. Equivalently, the measurable structure of $\mathscr{M}(X)$ is determined by the requirement that for any bounded measurable function on X, its coupling with M should be a measurable function in M. If X is standard, then $\mathscr{M}(X)$ is standard, too.

Observe that any link $\Lambda : X \dashrightarrow Y$ gives rise to a measurable map $\mathscr{M}(X) \to \mathscr{M}(Y)$, which we write as $M \mapsto M\Lambda$. Consequently, any projective system $\{V_i, \Lambda_i^j\}$ in $\mathscr{B}$ gives rise to the conventional projective limit of sets

$$\mathscr{M}_\infty := \varprojlim_I \mathscr{M}(V_i).$$

An element of $\mathscr{M}_\infty$ is called a *coherent family of measures*. By the very definition, it is a family of probability measures $\{M_i \in \mathscr{M}(V_i) : i \in I\}$ such that for any couple $i < j$ one has $M_j \Lambda_i^j = M_i$. (In the case of a transition function, Dynkin [35] calls elements of $\mathscr{M}_\infty$ *entrance laws*.)

If a limit object X exists then there is a canonical map

$$\mathscr{M}(X) \to \mathscr{M}_\infty.$$

From now on we will gradually narrow the setting of the formalism and will finally focus on the study of some concrete examples.

7.2 Projective Chains

Consider a particular case of a projective system where all spaces are discrete (finite or countably infinite) and the indices range over the set $\{1, 2, \ldots\}$ of natural numbers. Such a system is uniquely determined by the links Λ_N^{N+1}, $N = 1, 2, \ldots$:

$$V_1 \dashleftarrow V_2 \dashleftarrow \cdots \dashleftarrow V_N \dashleftarrow V_{N+1} \dashleftarrow \cdots . \tag{7.1}$$

Note that a link between two discrete spaces is simply a stochastic matrix, so that $\Lambda_N^{N+1} : V_{N+1} \dashrightarrow V_N$ is a stochastic matrix whose rows are parameterized by points of V_{N+1} and columns are parameterized by points of V_N:

$$\Lambda_N^{N+1} = [\Lambda_N^{N+1}(x, y)], \quad x \in V_{N+1},\ y \in V_N,$$

$$\Lambda_N^{N+1}(x, y) \geq 0 \text{ for every } x, y, \quad \sum_{y \in V_N} \Lambda_N^{N+1}(x, y) = 1 \text{ for every } x.$$

For arbitrary $N' > N$, the corresponding link $\Lambda_N^{N'} : V_{N'} \dashrightarrow V_N$ is a $V_{N'} \times V_N$ stochastic matrix, which factorizes into a product of stochastic matrices corresponding to couples of adjacent indices:

$$\Lambda_N^{N'} = \Lambda_{N'-1}^{N'} \cdots \Lambda_N^{N+1}.$$

We call such a projective system a *projective chain*. It gives rise to a chain of ordinary maps

$$\mathcal{M}(V_1) \leftarrow \mathcal{M}(V_2) \leftarrow \cdots \leftarrow \mathcal{M}(V_N) \leftarrow \mathcal{M}(V_{N+1}) \leftarrow \cdots . \tag{7.2}$$

Note that $\mathcal{M}(V_N)$ is a simplex whose vertices can be identified with the points of V_N, and the arrows are affine maps of simplices. In this situation a coherent family (element of $\mathcal{M}_\infty$) is a sequence $\{M_N \in \mathcal{M}(V_N) : N = 1, 2, \ldots\}$ such that

$$M_{N+1}\Lambda_N^{N+1} = M_N, \qquad N = 1, 2, \ldots .$$

Here we can interpret measures as row vectors, so that the left-hand side is the product of a row vector by a matrix. In more detail, the equation can be written as

$$\sum_{x \in V_{N+1}} M_{N+1}(x)\Lambda_N^{N+1}(x, y) = M_N(y), \qquad \forall y \in V_N.$$

Note that the set $\mathcal{M}_\infty$ may be empty, as the following simple example shows: take $V_N = \{N, N+1, N+2, \ldots\}$ and define Λ_N^{N+1} as the natural embedding $V_{N+1} \subset V_N$. In what follows we tacitly assume that $\mathcal{M}_\infty$ is nonempty. This holds automatically if all V_N are finite sets.

We may view $\mathscr{M}_\infty$ as a subset of the real vector space

$$L := \mathbb{R}^{V_1 \sqcup V_2 \sqcup V_3 \sqcup \cdots}.$$

Here the set $V_1 \sqcup V_2 \sqcup V_3 \sqcup \cdots$ is the disjoint union of V_Ns. Since this set is countable, the space L equipped with the product topology is locally convex and metrizable. Clearly, $\mathscr{M}_\infty$ is a convex Borel subset of L, hence a standard measurable space.

Let V_∞ be the set of extreme points of $\mathscr{M}_\infty$. We call V_∞ the *boundary* of the chain $\{V_N, \Lambda_N^{N+1}\}$.

Theorem 7.1 *If $\mathscr{M}_\infty$ is nonempty then the boundary $V_\infty \subset \mathscr{M}_\infty$ is a nonempty measurable subset (actually, a subset of type G_δ) of $\mathscr{M}_\infty$, and there is a natural bijection $\mathscr{M}_\infty \leftrightarrow \mathscr{M}(V_\infty)$, which is an isomorphism of measurable spaces.*

A proof based on Choquet's theorem is given in Olshanski [91, §9]; a much more general result is contained in Winkler [134, Chapter 4].

By the very definition of the boundary V_∞, it comes with canonical links

$$\Lambda_N^\infty : V_\infty \dashrightarrow V_N, \qquad N = 1, 2, \ldots.$$

Namely, given a point $\omega \in V_\infty \subset \mathscr{M}_\infty$, let $\{M_N\}$ stand for the corresponding sequence of measures; then, by definition,

$$\Lambda_N^\infty(\omega, x) = M_N(x), \qquad x \in V_N, \quad N = 1, 2, \ldots.$$

Here, to simplify the notation, we write $\Lambda_N^\infty(\omega, x)$ instead of $\Lambda_N^\infty(\omega, \{x\})$.

From the definition of Λ_N^∞ it follows that

$$\Lambda_{N+1}^\infty \Lambda_N^{N+1} = \Lambda_N^\infty, \qquad N = 1, 2, \ldots.$$

Now it is easy to see that the boundary V_∞ coincides with the categorical projective limit of the initial chain (7.1).

Remark 7.2 In the context of Theorem 7.1, assume we are given a standard measurable space X and links $\Lambda_N^X : X \dashrightarrow V_N$, $N = 1, 2, \ldots$, such that:

- $\Lambda_{N+1}^X \Lambda_N^{N+1} = \Lambda_N^X$ for all N;
- the induced map $\mathscr{M}(X) \to \mathscr{M}_\infty = \varprojlim \mathscr{M}(V_N)$ is a bijection.

Then X coincides with the boundary V_∞. Indeed, the maps $\mathscr{M}(X) \to \mathscr{M}_N$ are measurable, whence the map $\mathscr{M}(X) \to \mathscr{M}_\infty$ is measurable, too. Since $\mathscr{M}(X)$ is standard (because X is standard), the latter map is an isomorphism of measurable spaces (see, e.g., Parthasarathy [100, Chapter V, Theorem 2.4]) and the claim becomes obvious.

Remark 7.3 Theorem 7.1 immediately extends to the case of a projective system $\{V_i, \Lambda_i^j\}$, where all V_is are discrete spaces (finite or countable) and the directed index set I is countably generated; that is, contains a sequence $i(1) < i(2) < \cdots$ such that any $i \in I$ is majorated by indices $i(N)$ with N large enough. Indeed, it suffices to observe that the space $\varprojlim \mathscr{M}(V_{i(N)})$ does not depend on the choice of $\{i(N)\}$. Such a situation is examined in Borodin–Olshanski [23], where the index set I is the halfline $\mathbb{R}_{>0}$.

7.3 Graded Graphs

Definition 7.4 By a *graded graph* we mean a graph Γ with countably many vertices partitioned into *levels* enumerated by numbers $1, 2, \ldots$, and such that (below $|v|$ denotes the level of a vertex v):

- if two vertices v, v' are joined by an edge then $|v| - |v'| = \pm 1$;
- multiple edges between v and v' are allowed;
- each vertex v is joined with a least one vertex of level $|v| + 1$ (that is, there are no suspended vertices);
- if $|v| \geq 2$, then the set of vertices of level $|v| - 1$ joined with v is finite and nonempty.

This is a natural extension of the well-known notion of a *Bratteli diagram* (Bratteli [27]); the difference between the two notions is that a Bratteli diagram has finitely many vertices at each level, whereas our definition allows countable levels.

Sometimes it is convenient to slightly modify the above definition by adding to Γ a single vertex of level 0 joined by edges with all vertices of level 1.

The simplest nontrivial example of a graded graph is the Pascal graph $\mathbb{P}$ that we discussed in Chapter 5. A number of other examples can be found in Kerov [63] and also in Gnedin [43], Gnedin and Olshanski [46], Kingman [70].

Definition 7.5 Given a chain of finite or, more generally, compact groups embedded into each other,

$$G(1) \subset G(2) \subset \cdots \subset G(N-1) \subset G(N) \subset \cdots, \tag{7.3}$$

one constructs a graded graph $\Gamma = \Gamma(\{G(N)\})$, called the *branching graph* of the group chain (7.3), as follows. The vertices of level N are the labels of the equivalence classes of irreducible representations of $G(N)$. Choose a representation π_v for each vertex v. Two vertices v and u of levels N and $N-1$, respectively, are joined by m edges if π_u enters the decomposition of

$\pi_v \downarrow G(N-1)$ with multiplicity m, with the understanding that there are no edges if $m = 0$.

Of particular importance are the *Young graph* $\mathbb{Y}$ that we studied earlier, and the *Gelfand–Tsetlin graph* $\mathbb{GT}$; they are obtained from the chains of symmetric groups ($G(N) = S(N)$) and unitary groups ($G(N) = U(N)$), respectively.

Definition 7.6 By a *monotone path* in a graded graph Γ we mean a finite or infinite collection

$$v_1, e_{12}, v_2, e_{23}, v_3, \ldots$$

where $v_1, v_2, \ldots$ are vertices of Γ such that $|v_{i+1}| = |v_i| + 1$ and $e_{i,i+1}$ is an edge between v_i and v_{i+1}. Since we do not consider more general paths, the adjective "monotone" will often be omitted. If the graph is simple then every path is uniquely determined by its vertices, but when multiple edges occur it is necessary to specify which of the edges between every two consecutive vertices is selected.

Definition 7.7 Given a graded graph Γ, the *dimension* of a vertex v, denoted by $\dim v$, is defined as the number of all paths in Γ of length $|v| - 1$ starting at some vertex of level 1 and ending at v. Further, for an arbitrary vertex u with $|u| < |v|$, the *relative dimension* $\dim(u, v)$ is the number of monotone paths of length $|v| - |u|$ joining u to v. In particular, if $|u| = |v| - 1$, then $\dim(u, v)$ is the number of edges between u and v; let us denote it by $\varkappa(u, v)$.

For instance, in the case of the Pascal graph $\Gamma = \mathbb{P}$ the dimensions are binomial coefficients; see Chapter 5.

If Γ is the branching graph of a group chain, then $\dim v$ is the dimension of the corresponding representation π_v and $\dim(u, v)$ is the multiplicity of π_u in the decomposition of the representation π_v restricted to the subgroup $G(|u|) \subset G(|v|)$.

Obviously, for an arbitrary graded graph Γ one has

$$\dim v = \sum_{u\colon |u|=|v|-1} \dim u \ \dim(u, v).$$

Using this relation we assign to Γ a projective chain $\{V_N, \Lambda_N^{N+1}\}$, where $V_N = \Gamma_N$ consists of the vertices of level N and

$$\Lambda_N^{N+1}(v, u) = \frac{\dim u \dim(u, v)}{\dim v}, \qquad v \in \Gamma_{N+1}, \quad u \in \Gamma_N. \tag{7.4}$$

The boundary V_∞ of this chain is also referred to as the *boundary of the graph* Γ and denoted as $\partial\Gamma$.

More generally, for $N < N'$ we set

$$\Lambda_N^{N'} := \Lambda_{N'-1}^{N'} \cdots \Lambda_N^{N+1}.$$

Then

$$\Lambda_N^{N'}(v, u) = \frac{\dim u \dim(u, v)}{\dim v}, \qquad u \in \Gamma_N, \quad v \in V_{N'}. \tag{7.5}$$

Given a chain $\{G(N)\}$ of finite or compact groups, see (7.3), denote by G their union. *Characters* of G are defined according to Definition 1.7. (If G is a compact topological group, then we additionally require that the characters should be continuous functions on G; that is, restriction to every subgroup $G(N)$ is continuous.) As before, *extreme* characters are extreme points of the convex set of all characters.

Proposition 7.8 *If Γ is the branching graph of a chain* (7.3) *of finite or compact groups, then there is a natural bijection between the boundary $\partial\Gamma$ and the set of extreme characters of the group G.*

This is a generalization of Proposition 3.5, and the proof is the same.

7.4 Gibbs Measures

Let $X_1, X_2, \dots$ be an infinite sequence of nonempty sets, each of which is either finite or countable, and let $\varphi_{n,n+1} : X_{n+1} \to X_n$ be some surjective maps, $n = 1, 2, \dots$. Then we may form the projective limit set $X := \varprojlim X_n$, which is nonempty, because the maps are surjective. By the very definition of projective limit, for every n, there is a natural projection $\varphi_n : X \to X_n$. The inverse image $\varphi_n^{-1}(Y) \subset X$ of an arbitrary subset $Y \subset X_n$ under this projection is called a *cylinder set* of level n. We endow X with the sigma-algebra Σ generated by the cylinder sets of all levels. Next, let $\{\mu_n \in \mathscr{M}(X_n) : n = 1, 2, \dots\}$ be a sequence of probability measures, which are *consistent* with the maps $\varphi_{n,n+1}$ in the sense that $\varphi_{n,n+1}(\mu_{n+1}) = \mu_n$ for every $n = 1, 2, \dots$. We will need the following assertion:

Theorem 7.9 *Let X_n, $\varphi_{n,n+1}$, X, φ_n, Σ, and μ_n be as above. Then there exists a unique probability measure $\mu \in \mathscr{M}(X)$ defined on the sigma-algebra Σ, such that $\varphi_n(\mu) = \mu_n$ for every $n = 1, 2, \dots$*

The measure μ is called the *projective limit* of the sequence $\{\mu_n\}$.

Below we give a proof for the case when all sets X_n are finite, which suffices for our purposes. The case of countable sets is left to the reader as an exercise; see Exercise 7.10. Theorem 7.9 is a very particular case of *Bochner's theorem* (on projective systems of measures), which in turn is a generalization of the *Kolmogorov extension theorem*. References are given in the Notes to the present chapter.

Proof for the case of finite sets X_n For any given n, because the projection $\varphi_n : X \to X_n$ is surjective, the cylinder sets of level n form a sigma-algebra Σ_n, isomorphic to the sigma-algebra 2^{X_n} of all subsets of X_n. This makes it possible to interpret μ_n as a measure $\widetilde{\mu}_n$ on (X, Σ_n).

Next, observe that $\Sigma_1 \subseteq \Sigma_2 \subseteq \cdots$ and denote by Σ^0 the set-theoretic union $\bigcup \Sigma_n$. This is an algebra of sets, but, generally speaking, not a sigma-algebra. The consistency property of the sequence $\{\mu_n : n = 1, 2, \ldots\}$ means that, for each n, the restriction of $\widetilde{\mu}_{n+1}$ to Σ_n coincides with $\widetilde{\mu}_n$. Therefore, the sequence $\{\widetilde{\mu}_n\}$ gives rise to a finitely-additive set function $\widetilde{\mu}$ on Σ^0. We want to show that $\widetilde{\mu}$ admits a (unique) sigma-additive extension to Σ. According to a well-known criterion from the abstract measure theory (the Carathéodory theorem), to do this we have to prove that if $A_1 \supseteq A_2 \supseteq \cdots$ is a sequence of cylinder sets such that $\lim_i \widetilde{\mu}(A_i) > 0$, then the intersection $\bigcap A_i$ is nonempty.

Finally, observe that the space X has a natural topology in which the open sets are precisely the cylinder sets and their unions (Exercise 7.9). It is readily seen that X is a compact space and all cylinder sets are closed; hence they are compact, too. It follows that the intersection $\bigcap A_i$ is nonempty for *any* decreasing sequence of cylinder sets; we do not even need to use the assumption $\lim_i \widetilde{\mu}(A_i) > 0$. □

Now let, as above, Γ be a graded graph. Its *path space* $\mathscr{T} = \mathscr{T}(\Gamma)$ is defined as the set of all infinite paths starting at the first level. Likewise, denote by $\mathscr{T}_n$ the set of finite paths starting at the first level and ending at level n. Then we may write $\mathscr{T} = \varprojlim \mathscr{T}_n$. This enables one to introduce cylinder subsets of $\mathscr{T}$ and the corresponding sigma-algebra Σ, making $\mathscr{T}$ into a measurable space.

Definition 7.10 A probability measure on $\mathscr{T}$ is said to be a *Gibbs measure* if any two initial finite paths with the same endpoint are equiprobable. Equivalently, the measure of any cylinder set of the form $\varphi_n^{-1}(\tau) \subset \mathscr{T}$, where $\tau \in \mathscr{T}_n$, depends only on the endpoint of τ.

(In works of Vershik and Kerov, such measures are called *central measures*.)

As above, consider the projective chain $\{\Gamma_N, \Lambda_N^{N+1}\}$ associated with the graph Γ, so that Γ_N is the set of vertices of level $N = 1, 2, \ldots$.

Proposition 7.11 *There is a natural bijective correspondence $\{M_N\} \leftrightarrow \widetilde{M}$ between coherent systems of measures on Γ and Gibbs measures on $\mathscr{T}$.*

Proof Indeed, given a Gibbs measure $\widetilde{M}$, define for each N a probability measure $M_N \in \mathcal{M}(V_N)$ as follows: for any $v \in V_N$, $M_N(v)$ equals the probability that the infinite random path distributed according to P passes through v. The measures M_N are compatible with the links Λ_N^{N+1} by the very construction of these links. Therefore, the sequence (M_N) determines an element of $\mathcal{M}_\infty$. The inverse map, from $\mathcal{M}_\infty$ to Gibbs measures, is obtained by making use of Theorem 7.9. □

Together with Theorem 7.1 this implies:

Corollary 7.12 *There is a bijection between the Gibbs measures on the path space of* Γ *and the probability measures on the boundary* $\partial\Gamma$.

Note that the random paths distributed according to a Gibbs measure can be viewed as trajectories of a "Markov chain" with discrete time N that flows backwards from $+\infty$ to 0 and transition probabilities Λ_N^{N+1}. (It should be noted, however, that this is not a conventional Markov chain, because its state space varies with time.) Then probability measures on $\partial\Gamma$ turn into what can be called the *entrance laws* for that "Markov chain"; cf. Dynkin [34]. Thus, $\partial\Gamma$ plays the role of the entrance boundary, which is a justification of our use of the term "boundary". More generally, a similar interpretation can be given to the boundary of an arbitrary projective chain.

7.5 Examples of Path Spaces for Branching Graphs

For the Pascal graph $\Gamma = \mathbb{P}$, the path space can be identified with the space $\{0, 1\}^\infty$ of infinite binary sequences. Under this identification, the Gibbs measures are just the exchangeable measures on $\{0, 1\}^\infty$, and the claim of Corollary 7.12 turns into the classical de Finetti theorem: exchangeable probability measures on $\{0, 1\}^\infty$ are parameterized by probability measures on $[0, 1]$.

Consider the Young graph $\Gamma = \mathbb{Y}$. Recall that for a Young diagram $\lambda \in \mathbb{Y}$, a *standard Young tableau of shape* λ is a filling of the boxes of λ by numbers $1, 2, \ldots, |\lambda|$ in such a way that the numbers increase along each row from left to right and along each column from top to bottom.

Let us also define an *infinite Young diagram* as an infinite subset $\widetilde{\lambda} \subseteq \mathbb{Z}_{>0} \times \mathbb{Z}_{>0}$ such that if $(i, j) \in \widetilde{\lambda}$, then $\widetilde{\lambda}$ contains all pairs (i', j') with $i' \leq i$, $j' \leq j$. An *infinite standard tableau* of shape $\widetilde{\lambda}$ is an assignment of a positive integer to any pair $(i, j) \in \widetilde{\lambda}$ in a such a way that all positive integers are used, and they increase in both i and j. If we only pay attention to where the integers $1, 2, \ldots, n$ are located, we will observe a Young tableau whose shape is a Young diagram $\lambda \subset \widetilde{\lambda}$ with n boxes. Let us call this finite tableau the *n-truncation* of the original infinite one.

Clearly, the infinite paths in the Young graph are in one-to-one correspondence with the infinite Young tableaux. The initial finite parts of such a path are described by the various truncations of the corresponding tableau. The condition of a measure on infinite Young tableaux being Gibbs consists in the requirement that the probability of observing a prescribed truncation depends only on the shape of the truncation (and not on its filling).

The bijective correspondence between coherent systems on the Young graph $\mathbb{Y}$ and Gibbs measures on the path space of $\mathbb{Y}$ is employed in Chapter 12.

7.6 The Martin Boundary and the Vershik–Kerov Ergodic Theorem

Let $\{V_N, \Lambda_N^{N+1}\}$ be a projective chain and assume for simplicity that all sets V_N are finite. Let V be the disjoint union of the sets V_N and $\mathcal{F}$ denote the space of real-valued functions on V. To every $v \in V_N$ we assign a function $f_v \in \mathcal{F}$ as follows:

$$f_v(v') = \begin{cases} \Lambda_{N'}^{N}(v, v'), & \text{if } v' \in V_{N'} \text{ with } N' < N, \\ 1, & \text{if } v' = v, \\ 0, & \text{otherwise}. \end{cases}$$

In this way we get an embedding $V \to \mathcal{F}$, and to simplify the notation let us identify V with its image in $\mathcal{F}$. Next, equip $\mathcal{F}$ with the topology of pointwise convergence and take the closure of V in this topology. Denote this closure by $\bar{V}$.

Definition 7.13 By the *Martin boundary* V_{Martin} of the chain $\{V_N, \Lambda_N^{N+1}\}$ we mean the set difference $\bar{V} \setminus V$.

It is not hard to verify that $\bar{V}$ is a compact set. Thus, it is a compactification of V, which may be called the *Martin compactification.*

Observe that for every element $f \in V_{\text{Martin}}$ its restrictions to various sets V_N form a coherent system of distributions: indeed, this is easy to verify from the fact that f is a pointwise limit of a sequence $\{f_{v_i}\}$ where $v_i \in V_{N_i}$ and $N_1 < N_2 < \cdots$. (Note that here we substantially use the assumption of finiteness of the V_N.) Thus, we may view V_{Martin} as a subset of $\mathcal{M}_\infty$ of all coherent systems.

Theorem 7.14 *Let $\{V_N, \Lambda_N^{N+1}\}$ be a projective chain with finite sets V_N. As a subset of $\mathcal{M}_\infty$, the Martin boundary V_{Martin} contains the boundary V_∞.*

For a proof, see Kerov, Okounkov, and Olshanski [64].

The assertion of the theorem means that for any extreme coherent system $\{M_N\}$ there exists a sequence of points $\{v_i \in V_{N_i}\}$ with $N_i \to \infty$, such that for every N and every $v \in V_N$ one has

$$\lim_{i\to\infty} \Lambda_N^{N_i}(v_i, v) = M_N(v).$$

In particular, the above definitions and results are applicable to any graded graph Γ with finite levels, so that we may speak about its Martin boundary $\partial_{\mathrm{Martin}}\Gamma$ and Martin compactification. In the rest of the section we assume that Γ is such a graph.

Definition 7.15 Let Γ be a graded graph with finite levels.

(i) Given a sequence $\{v_i \in \Gamma_{N_i}\}$ of vertices with $N_i \to \infty$, let us say that $\{v_i \in \Gamma_{N_i}\}$ is *regular* if it converges to a point $\omega \in \partial_{\mathrm{Martin}}\Gamma$ of the Martin boundary in the Martin compactification of the graph. By the very definition of the Martin compactification, this means that for any fixed vertex $v \in \Gamma_N$, there exists a limit

$$\lim_{i\to\infty} \frac{\dim(v, v_i)}{\dim v_i}.$$

(ii) Likewise, a path $\tau \in \mathscr{T}(\Gamma)$ is said to *regular* if the sequence of its vertices is regular. Then we say that the path converges to the corresponding boundary point $\omega \in \partial_{\mathrm{Martin}}\Gamma$.

Remark 7.16 For the Young graph, the Martin boundary coincides with the boundary $\partial\mathbb{Y} = \Omega$, and the above definition of convergence to boundary points coincides with Definition 6.15; that is, a sequence $\lambda(n) \in \mathbb{Y}_n$ is regular if and only if the corresponding sequence $\frac{1}{n}\omega_{\lambda(n)}$ converges to a point of Ω. Indeed, this fact was established in the proof of Theorem 6.16.

Theorem 7.17 (Vershik–Kerov ergodic theorem) *Let Γ be a graded graph with finite levels, $\omega \in \partial\Gamma \subseteq \partial_{\mathrm{Martin}}\Gamma$ be an arbitrary point of the boundary, and $\widetilde{M}^{\omega}$ be the corresponding measure on the path space $\mathscr{T}(\Gamma)$.*

With respect to measure $\widetilde{M}^{\omega}$, almost all paths converge to ω.

For a proof we refer to Kerov [63, Chapter 1, Section 1].

It is easy to see that the set of regular paths is a Borel subset of the path space.

Corollary 7.18 *Let Γ be a graded graph with finite levels. Every Gibbs measure on the path space $\mathscr{T}(\Gamma)$ is concentrated on the subset of regular paths.*

This is a direct consequence of Theorems 7.17 and 7.1.

Corollary 7.19 (Strong law of large numbers for the Young graph) *Let $\omega \in \Omega$ be an arbitrary point and $\widetilde{M}^{\omega}$ be the corresponding Gibbs measure on the path space $\mathscr{T}(\mathbb{Y})$. For $\widetilde{M}^{\omega}$-almost every path $\tau = (\lambda(n) \in \mathbb{Y}_n)$, the scaled row and column lengths of diagrams $\lambda(n)$ have limits* (6.18), *where α_i and β_i are the coordinates of ω.*

Indeed, this follows from Theorem 7.17 and Remark 7.16.

7.7 Exercises

Exercise 7.1 Recall that the Pascal graph can be embedded into the Young graph: a vertex (k, l) of the former graph is identified with the hook Young diagram $\lambda_{k,l} := (k + \frac{1}{2} \mid l + \frac{1}{2})$. Check that this embedding agrees with the description of the boundaries, so that the boundary $\partial\mathbb{P}$ becomes a subset of the boundary $\partial\mathbb{Y}$.

Exercise 7.2 Let Γ be a graded graph. For $n = 0, 1, 2, \ldots$, let $\mathscr{T}_n$ denote the set of monotone paths in Γ starting at the root $\varnothing \in \Gamma_0$ and ending somewhere in the nth level. There are natural projections $\pi_{n,n-1} : \mathscr{T}_n \to \mathscr{T}_{n-1}$ which consist in removing the last edge. On the other hand, each $\mathscr{T}_n$ is equipped with an equivalence relation: for two paths $\tau, \tau' \in \mathscr{T}_n$ we write $\tau \sim_n \tau'$ if they have the same endpoint.

(a) The collection of the sets $\mathscr{T}_n$ together with projections $\mathscr{T}_n \to \mathscr{T}_{n-1}$ and equivalence relations $\sim_n$ allows one to reconstruct Γ. In particular, the vertices in Γ_n are identified with the equivalence classes in $\mathscr{T}_n$.

(b) Conversely, assume we are given a collection $\{\mathscr{T}_n\}_{n\geq 0}$ of sets together with projections $\pi_{n,n-1} : \mathscr{T}_n \to \mathscr{T}_{n-1}$ and an equivalence relation "$\sim_n$" in each $\mathscr{T}_n$. Assume also that $\mathscr{T}_0$ is a singleton. Show that these data come from a graded graph Γ if and only if the following conditions hold:

- All equivalence classes are finite.
- For any equivalence classes $u \subset \mathscr{T}_{n-1}$ and $v \subset \mathscr{T}_n$, and any element $\tau \in u$, the number of elements in the intersection $\pi^{-1}_{n,n-1}(\tau) \cap v$ depends only on u and v but not on the choice of τ in u. (Specifically, the number above is equal to the number of edges between u and v in the future graph.)
- The projections $\pi_{n,n-1}$ are surjective. (This guarantees the absence of suspended vertices.)

(c) The set of infinite paths in Γ started at $\varnothing$ can be identified with the projective limit $\varprojlim \mathscr{T}_n$.

(d) For the Young graph, the set $\mathscr{T}_n$ of paths is the set of Young tableaux with n boxes, the projection $\mathscr{T}_n \to \mathscr{T}_{n-1}$ consists in removing the box occupied by

n, and equivalence $\tau \sim_n \tau'$ means that the tableaux τ and τ' have the same shape.

Exercise 7.3 Recall that a *partition of a set* A is defined as a splitting of A into nonempty disjoint subsets called *blocks*. It should be emphasized that by definition, the blocks are assumed to be unordered. For $n = 1, 2, \dots$, let $\mathscr{T}_n$ be set of all partitions of the set $[n] := \{1, \dots, n\}$, and agree that $\mathscr{T}_0$ is a singleton. Observe that if A is a subset of B then any partition of B induces by restriction a partition of A. Applying this to $A = [n-1]$, $B = [n]$ we get projections $\mathscr{T}_n \to \mathscr{T}_{n-1}$. Two partitions of a set are said to be *equivalent* if they can be transformed to each other by a permutation of the underlying set. In particular, this defines an equivalence relation in each of the sets $\mathscr{T}_n$.

(a) Check that the data specified above obey the conditions of Exercise 7.2, so that they determine a branching graph. This graph is called the *Kingman graph*, we will denote it as $\mathbb{K}$ or, in more detail, as $(\mathbb{K}, \varkappa_{\mathbb{K}})$.

(b) The vertices of $\mathbb{K}$ can be identified with those of the Young graph $\mathbb{Y}$. Specifically, to a partition of the *set* $[n]$ one assigns the collection of the block lengths, which is a partition of the *number* n. Moreover, the edges in both graphs are the same, too. The difference is in the multiplicity function:

Show that for any couple $\mu \nearrow \lambda$ of Young diagrams, the "Kingman multiplicity" $\varkappa_{\mathbb{K}}(\mu, \lambda)$ equals the number of rows in λ having the same length as that of the row containing the box $\lambda \setminus \mu$.

(c) Show that the set $\varprojlim \mathscr{T}_n$ of infinite increasing paths in the Kingman graph can be identified with the set of all partitions of the infinite set $\{1, 2, \dots\}$.

(d) Show that the "Kingman dimension" is given by

$$\dim_{\mathbb{K}} \lambda = \frac{|\lambda|!}{\lambda_1! \lambda_2! \cdots}.$$

Exercise 7.4 Let us say that two graded graphs, $(\Gamma, \varkappa)$ and $(\Gamma', \varkappa')$ are *similar* if they have the same vertices and edges, and the multiplicity functions are conjugated by an appropriate strictly positive function ψ on the vertices:

$$\varkappa'(\mu, \lambda) = \frac{\psi(\mu)}{\psi(\lambda)} \varkappa(\mu, \lambda).$$

Show that for similar graded graphs, the respective sets of nonnegative normalized harmonic functions are isomorphic as convex sets. Furthermore, the respective boundaries are also essentially the same.

Exercise 7.5 Show that each of the following two modifications of $\varkappa_{\mathbb{K}}(\mu, \lambda)$ leads to a graded graph similar to the Kingman graph:

- $\varkappa(\mu, \lambda)$ equals the length of the row in λ containing the box $\lambda \setminus \mu$;
- $\varkappa(\mu, \lambda)$ equals the product mk, where k denotes the length of the row in λ containing the box $\lambda \setminus \mu$ while m is the number of all rows in λ of length k.

Exercise 7.6 Show that the graphs $\mathbb{K}$ and $\mathbb{Y}$ are not similar.

Exercise 7.7 A graded graph Γ is called *multiplicative* if there exists a unital $\mathbb{Z}_+$-graded $\mathbb{R}$-algebra $\mathscr{A} = \oplus \mathscr{A}_n$ together with a distinguished homogeneous basis $\{a_\lambda\}$ in $\mathscr{A}$ indexed by vertices $\lambda \in \Gamma$ such that the following conditions hold:

(1) If $\lambda \in \Gamma_n$ then the degree of the corresponding element a_λ equals n, so that $\dim \mathscr{A}_n = |\Gamma_n|$.
(2) The element $a_\varnothing$ is the unit element of $\mathscr{A}$.
(3) $\dim \mathscr{A}_1 = |\Gamma_1| = 1$. We will denote the unique vertex in Γ_1 as (1).
(4) For any $n \in \mathbb{Z}_+$ and any $\lambda \in \Gamma_n$,

$$a_\lambda \cdot a_{(1)} = \sum_\nu \varkappa(\lambda, \nu) a_\nu,$$

summed over the vertices $\nu \in \Gamma_{n+1}$ connected to λ by an edge, with coefficients equal to the formal edge multiplicities.

(a) Show that the Pascal graph is multiplicative. (Hint: take as $\mathscr{A}$ the subalgebra in $\mathbb{R}[x, y]$ formed by polynomials divisible by $a_{(1)} := x + y$.)
(b) Show that the Young graph $\mathbb{Y}$ is multiplicative – here $\mathscr{A}$ the algebra Sym of symmetric functions with the distinguished basis formed by the Schur functions s_λ.
(c) Show that the Kingman graph $\mathbb{K}$ is multiplicative: here again $\mathscr{A} = \text{Sym}$, but as the basis one has to take the monomial symmetric functions m_λ; see Chapter 1.

Exercise 7.8 The combinatorial notion of *composition* is similar to that of partition; the difference is that partitions are *unordered* collections while compositions are *ordered* ones. Again, there are two parallel notions: composition of a natural number n and composition of a set A. The former is defined as an ordered collection of strictly positive integers with sum n, and the latter is a splitting of A into disjoint nonempty blocks together with an ordering of the blocks. Repeating the definition of the graph $\mathbb{K}$ (see Exercise 7.3) with partitions replaced by compositions, one gets a branching graph called the *graph of compositions*.

Show that the dimension of vertices of this graph is given by the same expression as in the case of the Kingman graph (multimonomial coefficients), and compute the multiplicity function.

Exercise 7.9 Let $X = \varprojlim X_n$ be a projective limit of nonempty finite sets. Show that X is nonempty, too. Equip it with the topology in which the open sets are precisely the cylinder sets and their unions. Show that X is a compact Hausdorff space in this topology.

Exercise 7.10 Extend the proof of Theorem 7.9 given in the text to the case when the sets X_n are countable. (Hint: again, by the Carathéodory theorem, it suffices to show that if $A_1 \supseteq A_2 \supseteq \cdots$ is a sequence of sets from Σ^0 such that $\lim_n \widetilde{\mu}(A_n) > 0$, then $\bigcap A_n \neq \varnothing$. Passing to a subsequence of indices one may assume that $A_n \in \Sigma_n$. Show that there exists another sequence $\{B_n \in \Sigma_n : n = 1, 2, \dots\}$ such that $B_n \subseteq A_n$, the base $\varphi_n(B_n)$ of the cylinder B_n is a nonempty finite subset of X_n, and $B_1 \supseteq B_2 \supseteq \cdots$.)

7.8 Notes

A number of references have already been given in the main text; here are some additional comments.

The key ideas of the general formalism discussed in this chapter can be traced back to Dynkin's work on the entrance and exit boundaries of Markov processes, see [35] and references therein. Other important sources are Diaconis–Freedman [32] and Kerov's dissertation [63], which contains a survey of earlier works by him and Vershik. In our presentation we follow our papers [21], [24], where this formalism is applied to constructing Markov processes on limit objects.

About the Martin boundary for Markov chains, see, e.g., Sawyer's survey paper [108] and references therein; note, however, that Sawyer writes about the exit boundary while we are dealing with the entrance boundary.

The original version of the Kolmogorov extension theorem [71, Chapter III, Section 4] deals with probability measures on product spaces of the form $\mathbb{R}^{\mathscr{N}}$, where $\mathscr{N}$ is an arbitrary index set. This theorem and its refined versions are presented in many textbooks, see, e.g., Bogachev [5, Section 7.7] or Parthasarathy [100, Chapter V]. In Bochner's theorem (Bochner [6, Theorem 5.1.1], Parthasarathy [100, Chapter V]), product spaces are replaced by more general projective limit spaces.

Exercises 7.2–7.5 are taken from Kerov [63, Chapter 1, §§2–3]. The graph of compositions (Exercise 7.8) is studied in Gnedin [43]. The notion of multiplicative graph (Exercise 7.7) is due to Vershik and Kerov; see their paper [126].

Here is a list of works containing the computation of the boundaries of various graded graphs, which are somewhat related to our main examples (Pascal, Kingman, Young):

(a) Kerov [63, Chapter 1, §4] and Gnedin–Pitman [50] (Stirling triangles); Gnedin–Olshanski [46] (Euler triangle); Gnedin–Olshanski [47] (q-Pascal);

(b) Kingman [70] (the Kingman graph); Gnedin [43] (the graph of compositions, a suspension over the Kingman graph);

(c) Goodman–Kerov [51] and Gnedin–Kerov [44] (Young–Fibonacci, a curious relative of the Young graph); Gnedin–Olshanski [45] (the graph of zigzags, also known as subword order; its boundary is a suspension over that of the Young graph); Kerov–Okounkov–Olshanski [64] (the Young graph with formal Jack edge multiplicities); Borodin–Olshanski [23] (a graded poset with a continuous scale of levels, which serves as an intermediary between the Young graph and the Gelfand–Tsetlin graph).

A new formalism related to boundaries of graded graphs is developed in Vershik's papers [118], [119], [120].

PART TWO

UNITARY REPRESENTATIONS

8

Preliminaries and Gelfand Pairs

We start with a few well-known general definitions and facts concerning unitary representations.

Let G be an abstract group. A *unitary representation* of G in a complex Hilbert space H is a homomorphism of G into the group $U(H)$ of unitary operators in H. If T stands for the symbol of a representation, we denote by $T(g)$ the unitary operator corresponding to a group element g, and we often write $H(T)$ for the Hilbert space of T.

Two unitary representations T and T' of the same group are *equivalent* if there exists a surjective isometric map $H(T) \to H(T')$ transforming operators $T(g)$ to operators $T'(g)$. Equivalent representations are usually viewed as indistinguishable ones.

The *commutant* of T is the set of all bounded operators on $H(T)$ commuting with all operators $T(g)$, $g \in G$. The commutant is an algebra closed under passage to adjoint operator.

An *invariant subspace* of a unitary representation T is a closed subspace $H' \subset H(T)$ which is invariant under the action of all operators $T(g)$. Then the orthogonal complement to H' is an invariant subspace, too. The restriction of T to an invariant subspace H', denoted as $T\big|_{H'}$, gives rise to a *subrepresentation* of T. If T does not admit proper invariant subspaces then T is said to be *irreducible*.

Given a vector $\xi \in H(T)$, there exists a smallest invariant subspace in $H(T)$ containing ξ – this is the closure of the linear span of the orbit $\{T(g)\xi : g \in G\}$. This subspace is called the *cyclic span* of ξ. If it coincides with the whole space $H(T)$ (that is, the orbit is a total set) then ξ is called a *cyclic vector*. For a countable group G, if T is a unitary representation admitting a cyclic vector then $H(T)$ is separable. If T is irreducible then any nonzero vector is cyclic.

Proposition 8.1 (Analog of Schur's lemma) *A unitary representation T is irreducible if and only if its commutant reduces to scalar operators.*

Proof Assume T is reducible and let $H' \subset H(T)$ be a proper invariant subspace. The orthogonal projection onto H' is a nonscalar projection operator belonging to the commutant of T.

Conversely, if the commutant contains a nonscalar operator A then the commutant also contains the Hermitian operators $A + A^*$ and $i(A - A^*)$. At least one of them is nonscalar; let us denote it as B. Since B is nonscalar, its spectrum does not reduce to a singleton, therefore there exists a nontrivial spectral projection operator associated with B. This projection operator is contained in the commutant, too, because the spectral decomposition of B is invariant under conjugation by unitary operators $T(g)$, $g \in G$.

Thus, we have proved that T is reducible if and only if its commutant is not reduced to scalar operators, which is equivalent to the claim. □

Proposition 8.2 (Analog of Burnside's theorem) *Let T be a unitary representation of G,* End H *be the algebra of all bounded operators on $H = H(T)$, and $\mathfrak{A} \subset$* End H *be the subalgebra generated by the operators $T(g)$, $g \in G$.*

If T is irreducible then $\mathfrak{A}$ is dense in End H *in the strong operator topology.*

Proof Fix an arbitrary $k = 1, 2, \dots$ and consider the unitary representation T_k of the same group acting on the Hilbert space $H \otimes \mathbb{C}^k$; by definition, $T_k(g) = T(g) \otimes 1$.

Step 1. We claim that the commutant of T_k coincides with $1 \otimes \operatorname{End} \mathbb{C}^k$. Indeed, let $e_1, \dots, e_k$ be the canonical orthonormal basis of $\mathbb{C}^k$ and $P_1, \dots, P_k$ denote the one-dimensional projections onto $\mathbb{C}e_1, \dots, \mathbb{C}e_k$, respectively. Let A be an operator from the commutant of T_k. Then all operators of the form $P_i A P_j$ also belong to the commutant of T_k. On the other hand, these operators can be viewed as operators in H, and in such an interpretation they should belong to the commutant of T; hence they should be scalar operators, by virtue of Proposition 8.1. This just means that $A \in 1 \otimes \operatorname{End} \mathbb{C}^k$.

Step 2. We may identify $H \otimes \mathbb{C}^k$ with the direct sum of k copies of H. Let $\xi_1, \dots, \xi_k$ be linearly independent vectors in H and ξ be their direct sum, so that ξ may be viewed as a vector from $H \otimes \mathbb{C}^k$. Denote by H_ξ the closure of the subspace $(\mathfrak{A} \otimes 1)\xi$ in $H \otimes \mathbb{C}^k$. We claim that H_ξ coincides with the whole space $H \otimes \mathbb{C}^k$. Indeed, H_ξ is an invariant subspace of the representation T_k. By virtue of Step 1, the projection on H_ξ has the form $1 \otimes P$ where P is a projection in $\mathbb{C}^k$. Since ξ belongs to H_ξ and the vectors $\xi_1, \dots, \xi_k$ are linearly independent we conclude that $P = 1$.

Step 3. Finally, we observe that, by the very definition of the strong operator topology, Step 2 entails that $\mathfrak{A}$ is dense in End H. □

One of the fundamental ideas in representation theory is that many questions about unitary representations can be reduced to those about positive definite functions.

Recall the definition of a positive definite function on a group G (it has already been given in Chapter 1): a function $\varphi(g)$ on G is said to be *positive definite* if $\varphi(g^{-1}) = \overline{\varphi(g)}$ and, for any finite collection $g_1, \dots, g_n \in G$, the $n \times n$ matrix $[\varphi(g_j^{-1} g_i)]$ is nonnegative definite.

Proposition 8.3 (i) *Let T be a unitary representation of a group G and $\xi \in H(T)$ be a nonzero vector. Then the matrix element $\varphi(g) = (T(g)\xi, \xi)$ is a positive definite function.*

(ii) *Conversely, if $\varphi(g)$ is a nonzero positive definite function on G then there exists a unitary representation T with a cyclic vector ξ such that the corresponding matrix element coincides with φ. Moreover, such a couple (T, ξ) is unique within a natural equivalence: if (T', ξ') is another couple with the same property then there exists a surjective isometry $H(T) \to H(T')$ sending ξ to ξ' and transforming operators $T(g)$ to operators $T'(g)$.*

Thus, if ξ is a cyclic vector of T, then all the information about T is hidden in the positive definite function $\varphi(g) = (T(g)\xi, \xi)$.

Proof *Step 1.* Let X be an arbitrary set. A *positive definite kernel* on X is a complex-valued function $\psi(x, y)$ on $X \times X$ such that for any finite subset $X' \subseteq X$, the matrix $[\psi(x, y)]_{x,y \in X'}$ with rows and columns indexed by points of X' (enumerated in any order) is Hermitian and nonnegative definite. In particular, this implies that $\psi(x, y) = \overline{\psi(y, x)}$ for any $x, y \in X$.

We claim that a function $\psi(x, y)$ on $X \times X$ is a positive definite kernel if and only if there exists a map $x \mapsto \xi_x$ of the set X into a complex Hilbert space H such that ψ coincides with the Gram matrix of the family $\{\xi_x\}$:

$$(\xi_x, \xi_y) = \psi(x, y), \qquad \forall x, y \in X.$$

Indeed, one implication is easy – any Gram matrix is readily seen to be a positive definite kernel. Conversely, assuming ψ to be a positive definite kernel we construct the required Hilbert space H as follows:

Take the vector space V of formal finite linear combinations of the symbols v_x, $x \in X$, and equip it with the inner product

$$\left(\sum_x a_x v_x, \sum_y b_y v_y\right) = \sum_{x,y} a_x \bar{b}_y \psi(x, y), \qquad a_x, b_y \in \mathbb{C}.$$

Since ψ is positive definite, this inner product is nonnegative definite. Let $V_0 \subset V$ be its null space, then V/V_0 is a pre-Hilbert space, and we take as H its completion. The vectors ξ_x defined as the images of the vectors v_x under the composition of maps $V \to V/V_0 \to H$ have the desired property.

Step 2. If $(H', \{\xi'_x\})$ is another Hilbert space together with a system of vectors satisfying the same condition $(\xi'_x, \xi'_y) = \psi(x, y)$, then it is readily checked that the assignment $\xi_x \mapsto \xi'_x$ extends to an isometry $H \to H'$. Moreover, if the family $\{\xi'_x\}$ is total in H' then this isometry is surjective.

Step 3. Let us prove claim (i). Take $X = G$ and write g, h instead of x, y. By the very definition, a function $\varphi(g)$ on G is positive definite if and only if the associated function in two variables, $\psi(g, h) = \varphi(h^{-1}g)$, is a positive definite kernel. If φ is a matrix element, $\varphi(g) = (T(g)\xi, \xi)$, then ψ is the Gram matrix of the vectors $\xi_g := T(g)\xi$. Therefore, by virtue of Step 1, φ is positive definite.

Step 4. Finally, let us prove claim (ii). Given a positive definite function $\varphi(g)$, pass to the associated positive definite kernel $\psi(g, h)$. According to Step 1, we can realize ψ as the Gram matrix of a total system of vectors $\xi_g \in H$, $g \in G$. Observe that for any $g' \in G$, we have $\psi(g'g, g'h) = \psi(g, h)$. Applying Step 2, we see that there exists a unitary operator $T(g')$ in H such that $T(g')\xi_g = \xi_{g'g}$ for all $g \in G$. Moreover, such an operator is unique, which implies that the assignment $g' \mapsto T(g')$ is a representation. Obviously, the representation T together with the vector $\xi := \xi_e$ have the desired property. The uniqueness claim in (ii) also follows from Step 2. □

The argument used in Steps 1–2 above is called the *Gelfand–Naimark–Segal construction* or GNS construction, for short.

Denote by $\Phi(G)$ the set of all positive definite functions on G. This is a convex cone in the linear space of functions on G. For $\varphi', \varphi \in \Phi(G)$, write $\varphi' \preccurlyeq \varphi$ (in words, φ' is *dominated* by φ) if $\varphi - \varphi' \in \Phi(G)$.

Proposition 8.4 *Let T be a unitary representation of G with a cyclic vector ξ and $\varphi \in \Phi(G)$ be the corresponding matrix element. There is a bijective correspondence between functions $\varphi' \in \Phi(G)$ dominated by φ and Hermitian operators A in the commutant of T, such that $0 \le A \le 1$. This bijection is established by the relation*

$$\varphi'(g) = (AT(g)\xi, \xi), \qquad g \in G.$$

Proof The correspondence $A \mapsto \varphi'$ is immediate. Indeed, assume A is in the commutant and $0 \le A \le 1$. Observe that the operator $A^{1/2}$ also lies in the commutant and the function φ' as defined above coincides with the matrix element associated with the vector $\xi' = A^{1/2}\xi$. Therefore, φ' is positive definite. Since φ equals the sum of φ' and a similar function built from $1 - A$, we see that $\varphi' \preccurlyeq \varphi$.

To establish the converse correspondence, $\varphi' \mapsto A$, let us apply the GNS construction both to φ and φ', or rather to the associated positive definite kernels ψ and ψ', and denote by $(H, \{\xi_g\})$ and $(H', \{\xi'_g\})$ the resulting couples. Let V be the space introduced in Step 1 of the proof of Proposition 8.3. The two spaces H and H' arise from two different inner products, $(\cdot\,,\,\cdot)$ and $(\cdot\,,\,\cdot)'$, on one and the same space V. The assumption $\varphi' \preccurlyeq \varphi$ precisely means that $(v, v)' \le (v, v)$ for any $v \in V$. It follows that the assignment $\xi_g \mapsto \xi'_g$ extends to a contractive linear map $B : H \to H'$. Moreover, B is equivariant with respect to transformations of the systems $\{\xi_g\}$ and $\{\xi'_g\}$ resulting from the left shifts $g \mapsto g'g$. Consequently, the operator $A = B^*B$ acting in H has the desired property. □

Set

$$\Phi_1(G) = \{\varphi \in \Phi(G) : \varphi(e) = 1\}.$$

This is a convex set which serves as a base of the cone $\Phi(G)$.

Corollary 8.5 *Let T be a unitary representation of G with a cyclic unit vector ξ, and $\varphi \in \Phi_1(G)$ be the corresponding matrix element. Then T is irreducible if and only if φ is an extreme point of the convex set $\Phi_1(G)$.*

Let K be a subgroup of G. If T is a unitary representation of G, we will denote by $H(T)^K$ the subspace of K-invariant vectors in $H(T)$. Obviously, if $\xi \in H(T)^K$ then the corresponding matrix element is a K-biinvariant function; that is, it is invariant with respect to the two-sided action of the subgroup K on the group G. Conversely, we have:

Proposition 8.6 *Let φ be a nonzero K-biinvariant function from $\Phi(G)$ and (T, ξ) be the corresponding unitary representation with a cyclic vector. Then this vector is K-invariant.*

Proof Indeed, we have to prove that $T(k)\xi = \xi$ for any $k \in K$. Since ξ is a cyclic vector, it suffices to check that, for any $g \in G$, $(T(k)\xi, T(g^{-1})\xi)$ does not depend on k, which is obvious:

$$(T(k)\xi, T(g^{-1})\xi) = (T(gk)\xi, \xi) = \varphi(gk) = \varphi(g).$$

(At first glance, it might seem strange that we have used only the right K-invariance of $\varphi \in \Phi(G)$ and not the two-sided invariance, but due to the relation $\varphi(g^{-1}) = \overline{\varphi(g)}$ these two properties are equivalent.) □

We will denote by $\Phi(G/\!/K)$ the subset in $\Phi(G)$ formed by K-biinvariant functions.

Proposition 8.7 *Let φ and φ' be two functions from $\Phi(G)$ such that $\varphi' \preccurlyeq \varphi$. Then $\varphi \in \Phi(G//K)$ entails $\varphi' \in \Phi(G//K)$.*

Proof Take the cyclic representation (T, ξ) corresponding to φ. By Proposition 8.6, ξ is K-invariant. Then apply Proposition 8.4. □

Set $\Phi_1(G//K) = \Phi(G//K) \cap \Phi_1(G)$. Obviously, $\Phi_1(G//K)$ is a convex set.

Corollary 8.8 *A function $\varphi \in \Phi_1(G//K)$ is an extreme point in $\Phi(G)$ if and only if it is an extreme point in $\Phi_1(G//K)$.*

The connection between unitary representations and positive definite functions becomes especially nice for a special class of representations to be discussed now.

We consider first the case of finite groups G and finite-dimensional representations. Recall that all such representations are unitarizable, and in the sequel we will consider unitary representations only.

Definition 8.9 Let G be a finite group and K be its subgroup. The pair (G, K) is called a *Gelfand pair* if the algebra $\mathbb{C}[G//K]$ of K-biinvariant functions on G (with multiplication given by convolution) is commutative.

Proposition 8.10 *Let K be a subgroup in a finite group G.*

(G, K) is a Gelfand pair if and only if for any irreducible representation T of the group G, the subspace $H(T)^K$ has dimension 0 or 1.

Proof Let T be a unitary representation of G with $H(T)^K \neq \{0\}$. Let us extend T from the group G to the group algebra $\mathbb{C}[G]$. Denote by P the orthogonal projection in $H(T)$ with range $H(T)^K$; then $P = T(p)$, where p is an idempotent in $\mathbb{C}[G]$:

$$p = \frac{1}{|K|} \sum_{k \in K} k.$$

Therefore, for any $f \in \mathbb{C}[G]$, $PT(f)P = T(p * f * p)$. It follows that

$$PT(\mathbb{C}[G])P = T(\mathbb{C}[G//K]).$$

Assume now that T is irreducible. Then, by Burnside's theorem, $T(\mathbb{C}[G])$ coincides with the algebra $\mathrm{End}(H(T))$ of all operators in $H(T)$. Consequently, the operators from $PT(\mathbb{C}[G])P = T(\mathbb{C}[G//K])$ restricted to $H(T)^K$ exhaust the whole algebra $\mathrm{End}(H(T)^K)$.

If (G, K) is a Gelfand pair then, by commutativity of $\mathbb{C}[G//K]$, $\mathrm{End}(H(T)^K)$ is commutative, whence $\dim(H(T)^K) = 1$.

Conversely, if $\dim(H(T)^K)$ is at most 1 for any irreducible T then the above argument shows that $T(\mathbb{C}[G//K])$ is commutative. Since the direct sum of all irreducible Ts is a faithful representation of $\mathbb{C}[G]$, we see that the algebra $\mathbb{C}[G//K]$ is commutative, whence (G, K) is a Gelfand pair. □

Here is a convenient way of checking the commutativity property for $\mathbb{C}[G//K]$.

Proposition 8.11 *Let K be a subgroup in a finite group G and let $\sigma : G \to G$ be an anti-automorphism of G (i.e., $\sigma(gh) = \sigma(h)\sigma(g)$) such that $\sigma(K) = K$. Assume that for any $g \in G$, the elements g and $\sigma(g)$ belong to the same double (K, K)-coset of G. Then (G, K) is a Gelfand pair.*

Proof Since $\sigma(K) = K$, the map σ induces an anti-automorphism of $\mathbb{C}[G//K]$. On the other hand, σ leaves any element of $\mathbb{C}[G//K]$ invariant because $K\sigma(g)K = KgK$. Hence, $\mathbb{C}[G//K]$ is commutative. □

Corollary 8.12 *For any finite group K, the pair $(K \times K, \operatorname{diag} K)$, where*

$$\operatorname{diag} K = \{(k, k) \mid k \in K\} \subset K \times K,$$

is a Gelfand pair.

Proof The map $\sigma(k_1, k_2) = (k_2^{-1}, k_1^{-1})$ satisfies the hypotheses of Proposition 8.11. □

To handle infinite groups we adopt the following modification of Definition 8.9:

Definition 8.13 Let G be an arbitrary group (not necessarily finite), and let K be its subgroup. Then we say that (G, K) is a *Gelfand pair* if for any unitary representation T the operators $PT(g)P$, where P is the orthogonal projection on $H(T)^K$ and g ranges over G, commute with each other. That is,

$$PT(g_1)PT(g_2)P = PT(g_2)PT(g_1)P \qquad \forall g_1, g_2 \in G.$$

As follows from the proof of Proposition 8.10, when G is finite, Definition 8.13 is equivalent to Definition 8.9. From now on, the notion of Gelfand pair will be understood according to Definition 8.13.

Proposition 8.14 *Let (G, K) be a Gelfand pair. Then for any irreducible unitary representation T of G, the dimension of $H(T)^K$ is at most* 1.

Proof The argument is similar to that used in the proof of Proposition 8.10. The only difference is that instead of Burnside's theorem we apply its generalization stated in Proposition 8.2. □

As we will see now, to check the condition of Definition 8.13 it is not necessary to know the representations of G.

Proposition 8.15 *Let G be a group and K be its subgroup. Assume that G is the union of an ascending chain of subgroups $G(n)$ and K is the union of ascending chain of subgroups $K(n)$. Furthermore, assume that for any n, $K(n)$ is contained in $G(n)$ and $(G(n), K(n))$ is a Gelfand pair. Then (G, K) is a Gelfand pair, too.*

Proof Let T be a unitary representation of G and P be the orthogonal projection on $H(T)^K$. Let us take arbitrary $g_1, g_2 \in G$ and check the relation of Definition 8.13.

By assumption, there exists n such that $g_1, g_2 \in G(n)$. Since $(G(n), K(n))$ is assumed to be a Gelfand pair, we have

$$P_n T(g_1) P_n T(g_2) P_n = P_n T(g_2) P_n T(g_1) P_n,$$

where P_n is the orthogonal projection on $H(T)^{K(n)}$. As $n \to \infty$, P_n converge to P in the strong operator topology (i.e., $P_n v \to Pv$ for any $v \in H(T)$), because $H(T)^K = \bigcap_{n \geq 1} H(T)^{K(n)}$. Since the multiplication of operators is jointly continuous in the strong operator topology on the bounded subsets of $\operatorname{End}(H(T))$, the commutation relation above yields in the limit $n \to \infty$ the same relation with P_n replaced by P. □

Corollary 8.16 *If a group K is the union of an ascending chain of finite subgroups $K(n)$ then $(K \times K, \operatorname{diag} K)$ is a Gelfand pair.*

Proof By Corollary 8.12, $(K(n) \times K(n), \operatorname{diag} K(n))$ are Gelfand pairs; hence we may apply Proposition 8.15. □

Definition 8.17 By a *spherical representation* of a Gelfand pair (G, K) we mean a unitary representation T of G possessing a cyclic K-invariant vector ξ. Such a vector will be called a *spherical vector*, and the corresponding matrix element $\varphi(g) = (T(g)\xi, \xi)$ will be called a *spherical function* of T.

From now on we assume that the spherical vectors ξ are normalized by $\|\xi\| = 1$, then the corresponding spherical functions take value 1 at $e \in G$ and hence are elements of $\Phi_1(G//K)$.

By virtue of Proposition 8.14, if T is an *irreducible* spherical representation then it possesses a unique (within a scalar factor $\zeta \in \mathbb{C}$, $|\zeta| = 1$) spherical vector. It is worth noting that the arbitrariness in the choice of the scalar factor ζ does not affect the spherical function. Thus, for any irreducible spherical representation T we dispose with a *canonically determined* matrix element $\varphi \in \Phi_1(G//K)$ – the spherical function. By virtue of Proposition 8.3, φ is a complete invariant of T. According to Corollaries 8.5 and 8.8, the spherical functions of irreducible spherical representations can be characterized as the extreme points of the convex set $\Phi_1(G//K)$.

Here is a case when irreducible spherical representations are easily described:

Proposition 8.18 *Let K be a finite group. The irreducible spherical representations for the Gelfand pair $(K \times K, \operatorname{diag} K)$ are exactly those of the form $\pi \otimes \bar{\pi}$, where π is an arbitrary irreducible representation of K and $\bar{\pi}$ denotes the conjugate representation. Under the canonical isomorphism $H(\pi \otimes \bar{\pi}) = \operatorname{End}(H(\pi))$, the spherical vector ξ for such a representation is a scalar operator in $\operatorname{End}(H(\pi))$.*

Proof The irreducible representations of $K \times K$ are of the form $\pi \otimes \pi'$, where π and π' are arbitrary irreducible representations of K. Any vector ξ in $H(\pi \otimes \pi') = H(\pi) \otimes H(\pi')$ may be viewed as a linear operator from $H(\overline{\pi'})$ to $H(\pi)$. Such a vector ξ is diag K-invariant if and only if the corresponding operator commutes with the action of K or *intertwines* $\overline{\pi'}$ and π. Schur's lemma says that ξ must vanish unless $\overline{\pi'}$ is equivalent to π, in which case (setting $\overline{\pi'} = \pi$) the intertwining operator is a scalar operator. □

We conclude this chapter by relating the above results to the material of the earlier chapters.

Proposition 8.19 *Let K be an arbitrary group. There is an isomorphism of convex sets, $\chi \leftrightarrow \varphi$, between the characters of K (in the sense of Definition 1.7) and the spherical functions $\varphi \in \Phi_1((K \times K)// \operatorname{diag} K)$.*

Proof The correspondence $\chi \leftrightarrow \varphi$ is established by

$$\varphi(g_1, g_2) = \chi(g_2^{-1} g_1), \qquad g_1, g_2 \in K.$$

Centrality of χ is equivalent to diag K-biinvariance, as is seen from the formula

$$\varphi(h_1 g_1 h_2, h_1 g_2 h_2) = \chi(h_2^{-1} g_2^{-1} g_1 h_2), \qquad h_1, h_2 \in K.$$

Finally, assuming χ to be central, the positive-definiteness of φ is equivalent to the positive-definiteness of χ. Indeed, this is seen from the relation $\varphi(h^{-1}g) = \chi(l^{-1}k)$, where $g = (g_1, g_2)$ and $h = (h_1, h_2)$ are arbitrary elements of $K \times K$ and $k = g_1 g_2^{-1} \in K$, $l = h_1 h_2^{-1} \in K$. □

Let us now apply the above results to a concrete example that is important for us – the infinite symmetric group $S(\infty)$ introduced in Definition 3.1. Consider the group $S(\infty) \times S(\infty)$ which we will call the infinite *bisymmetric* group.

Theorem 8.20 (i) *The pair* (G, K), *where* G *is the infinite bisymmetric group* $S(\infty) \times S(\infty)$ *and* $K = \operatorname{diag} S(\infty)$, *is a Gelfand pair.*

(ii) *There is a natural bijective correspondence* $\omega \leftrightarrow T^\omega$ *between points* ω *of the Thoma simplex* Ω *and (equivalence classes of) irreducible spherical representations of* (G, K).

(iii) *The spherical function* φ^ω *of* T^ω *has the form*

$$\varphi^\omega(g) = \chi^\omega(g_2^{-1} g_1), \qquad g = (g_1, g_2) \in G,$$

where χ^ω *is the extreme character of* $S(\infty)$ *labeled by* $\omega \in \Omega$.

(iv) *Let* (T, ξ) *be an arbitrary, not necessarily irreducible, spherical representation of* (G, K) *and* $\varphi \in \Phi_1(G//K)$ *be its spherical function. There exists a unique probability Borel measure* P *on* Ω *such that*

$$\varphi(g) = \int_\Omega \varphi^\omega(g) P(d\omega), \qquad \forall g \in G. \tag{8.1}$$

Proof (i) This follows from Corollary 8.16.

(ii) We know that the irreducible spherical representations are parameterized by extreme points of the convex set $\Phi_1(G//K)$. By Proposition 8.19, this convex set is isomorphic to the convex set of characters of $S(\infty)$ which is described by Thoma's theorem.

(iii) This follows from the proof of Proposition 8.19.

(iv) In view of the correspondence

$$\text{spherical functions} \leftrightarrow \text{characters} \leftrightarrow \text{coherent systems},$$

this is simply a reformulation of the integral representation of coherent systems, see Theorem 3.12. □

8.1 Exercises

Exercise 8.1 Show that any irreducible unitary representation of a commutative group is one-dimensional.

Exercise 8.2 Let G be a finite group and T be its finite-dimensional representation. Show that T admits a cyclic vector if and only if the multiplicities of the irreducible subrepresentations in T do not exhaust their dimensions.

Exercise 8.3 Consider the group $G = \mathbb{Z}$, which is the simplest infinite group. Let $\mathbb{T} \subset \mathbb{C}$ be the unit circle with center at the origin, equipped with the Lebesgue measure, and let T be the unitary representation of $\mathbb{Z}$ in the Hilbert space $L^2(\mathbb{T})$ defined by

$$(T(n)f)(\zeta) = e^{in\zeta} f(\zeta), \qquad n \in \mathbb{Z}, \quad \zeta \in \mathbb{T}, \quad f \in L^2(\mathbb{T}).$$

(a) Show that T is equivalent to the regular representation of $\mathbb{Z}$ in $\ell^2(\mathbb{Z})$.

(b) The commutant of T consists of operators of multiplication by functions from $L^\infty(\mathbb{T})$.

(c) There is a 1–1 correspondence between invariant subspaces of T and measurable subsets of the unit circle considered modulo null subsets. Thus, T is highly reducible but does not admit nonzero *irreducible* subrepresentations.

Exercise 8.4 Let T be a unitary representation of a group G.

(a) If T is irreducible then the matrix elements corresponding to two given unit vectors, ξ_1 and ξ_2, coincide if and only if $\xi_1 = \zeta\xi_2$ where $\zeta \in \mathbb{C}$, $|\zeta| = 1$.

(b) Show by example that this claim may fail for reducible representations.

Exercise 8.5 Let $\varphi \in \Phi(G)$.

(a) Prove directly (without use of the associated representation) the inequality $|\varphi(g)| \le \varphi(e)$.

(b) Prove directly the inequality

$$|\varphi(g) - \varphi(h)|^2 \le 2\varphi(e)\left(\varphi(e) - \Re\varphi(gh^{-1})\right), \qquad g \in G, \quad h \in G.$$

(c) Prove directly the claim of Proposition 8.7.

Exercise 8.6 Assume (G, K) is a finite Gelfand pair.

(a) Let T be an irreducible spherical representation of (G, K) and χ be its character in the conventional sense, i.e., $\chi(g) = \operatorname{Tr}(T(g))$, $g \in G$. Prove that the spherical function of T can be written as

$$\varphi(g) = \frac{1}{|K|}\sum_{k \in K} \chi(gk) = \frac{1}{|K|}\sum_{k \in K} \chi(kg).$$

(b) Prove that the number of (equivalence classes of) irreducible spherical representations of (G, K) is equal to $|G//K|$, the number of double cosets modulo K.

(c) Prove that the algebra $\mathbb{C}[G//K]$ is isomorphic to the direct sum of $|G//K|$ copies of $\mathbb{C}$.

Exercise 8.7 Let K be a finite group. For any $\pi \in \widehat{K}$, the spherical function φ corresponding to the representation $\pi \otimes \bar{\pi}$ of the Gelfand pair $(K \times K, \operatorname{diag} K)$ has the form

$$\varphi(g_1, g_2) = \frac{\operatorname{Tr}\left(\pi(g_2^{-1} g_1)\right)}{\dim \pi}.$$

Exercise 8.8 Prove that the groups $G = S(k + l)$ and $K = S(k) \times S(l)$ form a Gelfand pair. (Here we realize G as the group of permutations of $\{1, 2, \ldots, k + l\}$ and K as the subgroup in G preserving the partition $\{1, \ldots, k\} \cup \{k + 1, \ldots, k + l\}$.) What is $|G//K|$ in this case?

Exercise 8.9 (a) Prove that $G = S(2n)$ realized as the group of permutations of $\{\pm 1, \ldots, \pm n\}$, and its hyperoctahedral subgroup

$$K = \{g \in G \mid g(-i) = -g(i),\ i = 1, \ldots, n\}$$

form a Gelfand pair.

(b) Let G and K be as in (a). Establish a one-to-one correspondence between double cosets modulo K and conjugacy classes in $S(n)$.

Exercise 8.10 Realize $G = S(2n + 1)$ as the group of permutations of the set $\{0, \pm 1, \ldots, \pm n\}$ and let $K \subset S(2n) \subset S(2n + 1)$ be the same subgroup as in (a). Prove that (G, K) is a Gelfand pair.

Exercise 8.11 Let L be a finite group and let K be its subgroup.

(a) Consider the following condition on L and K:

(*) The restriction of any irreducible representation π of the group L to the subgroup K is multiplicity free; that is, no irreducible representation of K enters $\pi\big|_K$ more than once.

For instance, (*) holds for $L = S(n+1)$ and $K = S(n)$ by virtue of Young's branching rule; see Proposition 1.4.

Prove that (*) is satisfied if and only if the pair $(L \times K, \operatorname{diag} K)$ is a Gelfand pair. This is a generalization of Corollary 8.12.

(b) Define a *K-conjugacy class in L* as a subset in L of the form $\{klk^{-1}\}$ where $l \in L$ is fixed while k ranges over K. Let us say that $f \in \mathbb{C}[L]$ is a *K-central element* if f is constant on the K-conjugacy classes.

Observe that the subspace in the group algebra $\mathbb{C}[L]$ formed by the K-central elements is a subalgebra. Prove that Condition (*) in (a) is equivalent to commutativity of this subalgebra.

(c) Prove directly that the subalgebra of $S(n)$-central elements in $\mathbb{C}[S(n+1)]$ is commutative. (Hint: apply Proposition 8.11.)

(d) Assume L and K satisfy Condition (*). Prove that the irreducible spherical representations of $(L \times K, \operatorname{diag} K)$ are precisely those of the form $\pi \otimes \bar{\sigma}$, where $\pi \in \widehat{L}$ and $\sigma \in \widehat{K}$ are such that σ enters $\pi\big|_K$. Prove also that the spherical function of $\pi \otimes \bar{\sigma}$ has the form

$$\varphi(l, k) = \psi_{\pi,\sigma}(lk^{-1}), \qquad l \in L, \quad k \in K,$$

with

$$\psi_{\pi,\sigma}(l) := \frac{1}{|K|} \sum_{k \in K} \chi^{\pi}(lk)\chi^{\sigma}(k^{-1}), \qquad l \in L,$$

where χ^π and χ^σ denote, respectively, the characters of π and σ in the conventional sense. Verify that in the case $K = L$ the result is the same as in Exercise 8.7.

(e) Assume L and K satisfy Condition (*). Prove that the number of irreducible spherical representations of $(L \times K, \operatorname{diag} K)$ equals the number of K-conjugacy classes in L.

(f) Check directly the equality in (e) for $L = S(n+1)$, $K = S(n)$. (An explicit computation of the spherical functions in this case is a nontrivial problem; it was solved by Strahov [114].)

8.2 Notes

For generalities on unitary representations see, e.g., Dixmier [33]. The literature on Gelfand pairs is immense; finite Gelfand pairs are considered in the monograph [28] by Ceccherini-Silberstein, Scarabotti, and Tolli. Definition 8.13 is taken from Olshanski [87]. Neretin's recent paper [76] provides a surprising example of a Gelfand pair (G, K) in which G is an infinite-dimensional group and K is a finite-dimensional noncompact Lie group. The observation stated as Proposition 8.19 was made in Olshanski [84]. In a wider context, a correspondence between certain factor representations of $S(\infty)$ and certain irreducible representations of $S(\infty) \times S(\infty)$ is established in Vershik–Nessonov [127].

9

Classification of General Spherical Type Representations

The main result of this chapter is Theorem 9.2. It is substantially employed in Chapter 12; there, we apply it to reduce Theorem 12.1 to Theorem 12.2.

Let (G, K) be a Gelfand pair. Recall that we have defined general spherical representations of (G, K) as *couples* (T, ξ), where ξ is a cyclic K-invariant unit vector (Definition 8.17). With this understanding of spherical representations, their classification is reduced to that of spherical functions $\varphi \in \Phi_1(G//K)$, and in the concrete case of interest for us it is afforded by Theorem 8.20 (iv).

However, the situation changes if one defines a spherical representation in a slightly different way:

Definition 9.1 Let (G, K) be a Gelfand pair and T be a unitary representation of G. Let us say that T is a *spherical type representation* if it possesses a cyclic K-invariant vector.

(Warning: "spherical type representation" is not a conventional term; we have invented it just to avoid confusion with Definition 8.17.)

The difference between this definition and Definition 8.17 is that now we do not fix a cyclic K-invariant vector, we only require its existence. The main consequence is that the notion of equivalence changes: equivalence of couples (T, ξ) is not the same as equivalence of simply representations T. Of course, for irreducible representations there is essentially no difference, because then the vector ξ is unique within a scalar factor. But this is not the case for reducible spherical type representations. If we want to classify them within usual equivalence, we are led to the following problem: under which condition two functions from $\Phi_1(G//K)$ determine (via the GNS construction) equivalent unitary representations?

In the present chapter we solve the problem in the concrete case of the bisymmetric group.

Let us recall a few necessary notions. First of all, it is worth noting that all measures on Thoma's simplex Ω under consideration are assumed to be Borel measures. Here we have in mind the measurable (Borel) structure in the space Ω generated by its topology. Recall that Ω is a nice topological space (metrizable and separable).

- Two measures are *equivalent* if they have the same null sets (that is, subsets of measure 0).
- A measure P' is said to be *absolutely continuous* with respect to another measure P if each null set for P is also a null set for P'. This definition depends on the respective equivalence classes only, which gives meaning to the expression that one class is absolutely continuous with respect to another one.
- Two measures P and P' are *disjoint* if there is no nonzero measure P'' which is absolutely continuous with respect to both P and P'. Equivalently, there exist two disjoint Borel subsets supporting P and P', respectively.
- Likewise, two unitary representations T and T' of one and the same group are called *disjoint* if they do not have equivalent subrepresentations.

  Recall also an important fact called the *Radon–Nikodým theorem*:
- *A measure P' is absolutely continuous with respect to a measure P if and only if P' can be written as a product fP where f is a nonnegative Borel function.* This function f is called the *Radon–Nikodým derivative* of P' with respect to P; it is unique within adding a function concentrated on a P-null subset. In particular, P and P' are equivalent if the above function f can be chosen to be nonvanishing.

Finally, in the context of Theorem 8.20(iv), let us say that formula (8.1) gives the *spectral decomposition* of φ and P is the *spectral measure* of φ.

Theorem 9.2 *Consider the Gelfand pair*

$$G = S(\infty) \times S(\infty), \qquad K = \operatorname{diag} S(\infty).$$

Let T and T' be two spherical type representations of (G, K) in the sense of Definition 9.1, φ be a spherical function of T, φ' be a spherical function of T', and P and P' be the corresponding spectral measures.

(i) *T is equivalent to T' if and only if P is equivalent to P'. Thus, equivalence classes of spherical type unitary representations of our pair (G, K) are in a one-to-one correspondence with equivalence classes of probability measures on Thoma's simplex Ω.*

(ii) *T' can be realized as a subrepresentation of T if and only if P' is absolutely continuous with respect to P.*

(iii) *In particular, T and T' are disjoint if and only if P and P' are disjoint.*

Our basic instrument for proving Theorem 9.2 is decomposition of a reducible spherical type representation into a direct integral of irreducible ones, which is parallel to the spectral decomposition of spherical functions. We proceed to a description of this construction.

It is easily verified that there exist precisely two one-dimensional spherical representations of G, the trivial and the sign ones, which send an element $g = (g_1, g_2) \in G$ respectively to 1 and to ± 1 (according to parity of the permutation $g_1 g_2^{-1}$). Denote by ω_1 and ω_{sgn} the corresponding points in Thoma's simplex:

$$\omega_1 = ((1, 0, 0, \ldots), \ (0, 0, \ldots)), \quad \omega_{\text{sgn}} = ((0, 0, \ldots), \ (1, 0, 0, \ldots)).$$

For all $\omega \in \Omega \setminus \{\omega_1, \omega_{\text{sgn}}\}$ the representations T^ω are infinite-dimensional and hence can be realized in one and the same separable Hilbert space (separability follows from the fact that G is a countable group).

Proposition 9.3 *Fix a separable Hilbert space E with a distinguished orthonormal basis $e_1, e_2, \ldots$. Assume ω ranges over $\Omega \setminus \{\omega_1, \omega_{\text{sgn}}\}$.*

It is possible to realize all representations T^ω together in the space E in such a way that e_1 be the spherical vector and for any fixed $g \in G$, the operator-valued function $\omega \mapsto T^\omega(g)$ be a Borel map.

By definition, the last property means that for any fixed vectors $e, f \in E$, the matrix element $(T(g)e, f)$ is a Borel function in ω.

Proof We assume ω ranges over $\Omega \setminus \{\omega_1, \omega_{\text{sgn}}\}$. Let $H^\omega = H(T^\omega)$ and $\xi^\omega \in H^\omega$ be the spherical vector, $\|\xi^\omega\| = 1$. The argument relies only on the following two facts: first, the spherical function $\varphi^\omega(g) = (T^\omega(g)\xi^\omega, \xi^\omega)$ depends continuously on the parameter ω (actually, Borel dependence would also suffice) and, second, ξ^ω is a cyclic vector in H^ω for any ω.

Using these two properties we will describe now a procedure of identification of the representation space H^ω with the model space E.

Let us enumerate the elements of the group G into a sequence $g_1, g_2, \ldots$ such that $g_1 = e$. For any ω, the vectors $\xi_i^\omega := T^\omega(g_i)\xi^\omega$ form a total system of vectors in H^ω. Applying the Gram–Schmidt orthogonalization process to this system we obtain an orthonormal basis in H^ω, say, $e_1^\omega, e_2^\omega, \ldots$. Then we identify H^ω with E by setting $e_i^\omega \to e_i$, $i = 1, 2, \ldots$. We claim that the resulting realization of T^ω in E has the required properties.

Obviously, $e_1 \in E$ becomes the spherical vector, because $g_1 = e$ implies $e_1^\omega = \xi^\omega$.

To prove the Borel dependence on ω it suffices to check that for any fixed $g \in G$ and any fixed indices k, l, the matrix element $(T^\omega(g)e_k^\omega, e_l^\omega)$ is a Borel function in ω.

If we were guaranteed that the vectors $\xi_1^\omega, \xi_2^\omega, \dots$ are linearly independent, we could conclude that they are transformed to the vectors $e_1^\omega, e_2^\omega, \dots$ by a triangular linear transformation,

$$e_n^\omega = c_{n1}^\omega \xi_1^\omega + \cdots + c_{nn}^\omega \xi_n^\omega, \qquad c_{nn}^\omega > 0, \tag{9.1}$$

where the coefficients $c_{n1}^\omega, \dots, c_{nn}^\omega$ are uniquely determined by the $n \times n$ Gram matrix $\mathscr{G}_n^\omega = [(\xi_i^\omega, \xi_j^\omega)]_{1 \le i,j \le n}$ and depend continuously on its entries. Since $(\xi_i^\omega, \xi_j^\omega) = \varphi^\omega(g_j^{-1} g_i)$, this would imply that these coefficients depend continuously on ω and hence any matrix coefficient $(T^\omega(g) e_k^\omega, e_l^\omega)$ is a continuous function in ω.

Actually, the situation may be slightly more involved due to possible linear dependence of the initial vectors – it may well happen that for some indices n, the vector ξ_n^ω is linearly expressed through $\xi_1^\omega, \dots, \xi_{n-1}^\omega$, which holds precisely when the rank of the Gram matrix $\mathscr{G}_n^\omega$ equals the rank of the matrix $\mathscr{G}_{n-1}^\omega$. Such redundant vectors should be removed from the orthogonalization process. Denoting by $\nu(1) = 1 < \nu(2) < \cdots$ the indices of the remaining vectors we still can claim the validity of (9.1), where we only have to substitute $\xi_{\nu(i)}^\omega$ instead of ξ_i^ω.

For any fixed N, we can split $\Omega \setminus \{\omega_1, \omega_{\mathrm{sgn}}\}$ into a disjoint union of subsets,

$$\Omega \setminus \{\omega_1, \omega_{\mathrm{sgn}}\} = \bigcup_{n_1 < \cdots < n_N} X(n_1, \dots, n_N)$$

where

$$X(n_1, \dots, n_N) = \{\omega : \nu(1) = n_1, \dots, \nu(N) = n_N\}.$$

This makes sense because the subsequence $\nu(1), \nu(2), \dots$ depends on ω. Observe that each subset $X(n_1, \dots, n_N)$ is singled out by a finite system of equalities of the form $f(\omega) = 0$ and inequalities of the form $g(\omega) \ne 0$ with certain continuous functions f or g, hence $X(n_1, \dots, n_N)$ are Borel sets. On every such subset, the first N vectors $e_1^\omega, \dots, e_N^\omega$ are expressed through a finite number of initial vectors $\xi_1^\omega, \xi_2^\omega, \dots$ with coefficients depending continuously on ω.

Finally, we choose $N = \max(k, l)$. Then, according to what has been said above, on each of the Borel subsets $X(n_1, \dots, n_N)$, the matrix element $(T^\omega(g) e_k^\omega, e_l^\omega)$ is a continuous function in ω for any fixed $g \in G$, which completes the proof. □

We have shown that one can attach to any point ω distinct from ω_1 and ω_{sgn} a concrete realization of the representation T^ω in the "model space" E, with sufficiently good ("non-pathological") dependence on ω. We can complete the picture by attaching to each of the excluded points ω_1 and ω_{sgn} the one-dimensional space $E_1 := \mathbb{C}e_1 \subset E$, where we realize the trivial and sign representations, respectively.

Let $\mathscr{E}$ denote the vector space of all Borel vector-valued functions $f : \Omega \to E$ such that $f(\omega_1) \in E_1$ and $f(\omega_{\text{sgn}}) \in E_1$. For any $f \in \mathscr{E}$, the scalar-valued function $\|f(\omega)\|$ is also Borel. We define a natural linear action of the group G in $\mathscr{E}$ by setting

$$(T(g)f)(\omega) = T^\omega(g)f(\omega), \qquad g \in G, \quad \omega \in \Omega.$$

Assume we are given a Borel probability measure P on Ω. Let us say that two functions from $\mathscr{E}$ are *equivalent* if they differ from each other on a P-null subset only. Let $L^2(P, \mathscr{E})$ stand for the set of equivalence classes of functions $f \in \mathscr{E}$ such that

$$\|f\|_P^2 := \int_\Omega \|f(\omega)\|^2 P(d\omega) < \infty.$$

This is a separable Hilbert space with the inner product

$$(f, f')_P = \int_\Omega (f(\omega), f'(\omega)) P(d\omega).$$

The linear operators $T(g) : \mathscr{E} \to \mathscr{E}$ defined above induce a unitary representation of the group G in $L^2(P, \mathscr{E})$. This representation, which we will denote as T_P, is called the *direct integral* of the representations T^ω with respect to the measure P. If P is a purely atomic measure, the direct integral is reduced to a direct sum decomposition.

Let $L^\infty(P)$ denote the space of equivalence classes of essentially bounded complex-valued Borel functions on Ω; again, we do not distinguish functions which differ from each other on a P-null subset. The operator of multiplication by a function from $L^\infty(P)$ is clearly a bounded operator in $L^2(P, \mathscr{E})$ commuting with the representation T_P.

Proposition 9.4 *The operators of multiplication by functions from $L^\infty(P)$ exhaust the whole commutant of T_P.*

Proof Let us abbreviate $H_P = L^2(P, \mathscr{E})$ and denote, as usual, by H_P^K the subspace of K-invariant vectors. Obviously, $H_P^K \subset H_P$ consists of functions with values in the one-dimensional subspace E_1. Thus, H_P^K can be identified with the Hilbert space $L^2(P)$ of square integrable complex-valued functions. Let Q be the projection $H_P \to H_P^K$. For any $g \in G$, the operator $QT_P(g)Q$, which we regard as an operator in $L^2(P)$, is the operator of multiplication by the function $F_g(\omega) := \varphi^\omega(g)$.

Step 1. Observe that the functions $F_g(\omega)$ just defined are precisely the products of the functions of the form $p_k^\circ(\omega)$ that were introduced in Chapter 3.

Indeed, write $g = (g_1, g_2)$ and denote by $\rho = (\rho_1 \geq \rho_2 \geq \cdots \geq \rho_\ell \geq 2)$ the cycle structure of $g_2^{-1} g_1$; then

$$F_g(\omega) = \chi_\rho^\omega = p_{\rho_1}^\circ(\omega) \cdots p_{\rho_\ell}^\circ(\omega).$$

We know that the functions $p_k^\circ(\omega)$ are continuous functions separating points of the compact space Ω (Proposition 3.9). This implies that any operator in $L^2(P)$ commuting with multiplications by these functions must be itself multiplication by a function from $L^\infty(P)$.

Step 2. On the other hand, we claim that the closed linear span of the subspaces $T_P(g)H_P^K$ is the whole space H_P. Indeed, let $f \in H_P$ be an arbitrary vector orthogonal to all these subspaces. That is, $T_P(g)f$ is orthogonal to H_P^K for all $g \in G$. This is equivalent to saying that for any g, the vector $T^\omega(g)f(\omega) \in E$ is orthogonal to e_1 for all ω outside a P-null subset. Since the group G is countable, such a negligible subset can be chosen simultaneously for all $g \in G$. Since T^ω are irreducible representations, we conclude that $f(\omega)$ vanishes P-almost everywhere, which proves the above claim.

Step 3. Let $A \in \operatorname{End} H_P$ be in the commutant of T_P. It is an easy abstract fact that A must commute with Q. Consequently, QAQ commutes with the operators of the form $QT_P(g)Q$, $g \in G$. Regard all these operators as operators in the space $H_P^K = L^2(\Omega, P)$. Recall that $QT_P(g)Q$ acts in this space as multiplication by $F_g(\omega)$. By virtue of Step 1, QAQ must coincide with the operator of multiplication by a certain function from $L^\infty(P)$. Denote the latter operator by M. Since both A and M commute with Q, this can be written as $(A - M)Q = 0$. Multiplying by $T_P(g)$ on the left we get

$$(A - M)T_P(g)Q = T_P(g)(A - M)Q = 0 \qquad \forall g \in G.$$

By virtue of Step 2, this implies $A = M$. □

Corollary 9.5 *Each invariant subspace of the representation T_P consists precisely of functions supported by a fixed Borel subset in Ω, which is determined uniquely, within a P-null subset.*

Let $\xi_P(\omega)$ be the function identically equal to e_1. Obviously, it is a unit vector from H_P^K.

Corollary 9.6 *ξ_P is a cyclic vector, so that T_P is a spherical representation.*

Proof Indeed, since $\xi_P(\omega)$ is a nonvanishing function, it is not contained in any proper invariant subspace, as follows from the previous corollary. □

Let $\varphi_P(g)$ denote the spherical function corresponding to ξ_P. Obviously, we have

$$\varphi_P(g) = (T_P(g)\xi_P, \xi_P) = \int_\Omega \varphi^\omega(g) P(d\omega), \qquad g \in G.$$

Proof of Theorem 9.2 (i) Any spherical function can be written as φ_P with an appropriate probability measure P (Theorem 8.20). It follows that any spherical representation is equivalent to one of the representations T_P. The question about equivalence of spherical representations is thus reduced to equivalence of representations of the form T_P. This in turn is reduced to the following question: given P, what spherical functions can be obtained from cyclic vectors in H_P^K?

Recall that the generic form of a vector from H_P^K is $\eta(\omega) = f(\omega)e_1$ where $f \in L^2(P)$. The corresponding spherical function is $\varphi_{P'}$ where $P' = |f|^2 P$. By Corollary 9.5, η is a cyclic vector if and only if $\eta(\omega) \neq 0$ almost everywhere with respect to P, which is precisely the same as saying that P' is equivalent to P.

(ii) The argument is similar. Given two probability measures P and P', we have to decide when $T_{P'}$ can be realized as a subrepepresentation of T_P. This is reduced to the question of when there exists a unit vector in H_P^K with the matrix element equal to $\varphi_{P'}$. As above, we conclude that this happens precisely when P' can be written in the form $|f|^2 P$ which means that P' is absolutely continuous with respect to P.

(iii) The argument is similar. □

9.1 Notes

For general facts regarding direct integral decomposition of unitary representations, see Dixmier [33].

10

Realization of Irreducible Spherical Representations of $(S(\infty) \times S(\infty), \mathrm{diag}S(\infty))$

By virtue of Theorem 8.20, irreducible unitary spherical representations of the Gelfand pair $(S(\infty) \times S(\infty), \mathrm{diag}S(\infty))$ are parameterized by points $\omega = (\alpha, \beta)$ of the Thoma simplex Ω. Let us denote these representations as T^{ω} or $T^{\alpha,\beta}$. Let $\underline{0}$ stand for the infinite sequence of 0s.

Proposition 10.1 *The biregular representation T of $S(\infty) \times S(\infty)$ in $\ell^2(S(\infty))$ defined by*

$$(T(g,h)f)(x) = f(g^{-1}xh), \qquad g,h,x \in S(\infty), \quad f \in \ell^2(S(\infty)),$$

is the irreducible spherical representation $T^{\underline{0},\underline{0}}$ that corresponds to the point $(\alpha, \beta) = (\underline{0}, \underline{0})$ of Ω.

Proof Taking $\xi \in \ell^2(S(\infty))$ to be the delta-function at the unit element we immediately see that ξ is $\mathrm{diag}S(\infty)$-invariant, and that the character corresponding to the spherical function associated with ξ is also the delta-function at the unit element of $S(\infty)$. Since by Thoma's theorem this is exactly the extreme character corresponding to $\alpha = \beta = \underline{0}$, it remains to show that ξ is cyclic, which is obvious. □

In fact, on irreducibility of biregular representations, the following more general statement is true:

Proposition 10.2 *Let K be a discrete group, all of whose conjugacy classes, except for the trivial one consisting of the unit element, are infinite. Then the biregular representation of $K \times K$ in $\ell^2(K)$ is irreducible.*

Proof If the biregular representation is a direct sum of two other ones, then the projection of the cyclic vector $\xi = \delta_e$ on either invariant subspace must be an element of $\ell^2(K)$ invariant with respect to diag K-action. But ξ is the only such vector because the nontrivial conjugacy classes of K are infinite. □

The main goal of this chapter is to construct the irreducible spherical representations of $(S(\infty) \times S(\infty), \operatorname{diag} S(\infty))$ that correspond to arbitrary $(\alpha, \beta) \in \Omega$.

In what follows we will need the notion of tensor product of an infinite number of Hilbert spaces.

Recall first the definition of a finite tensor product of Hilbert spaces. Let $H_1, H_2, \ldots, H_n$ be Hilbert spaces. Take first the algebraic tensor product of these spaces. It is a pre-Hilbert space with the inner product defined on polyvectors by

$$(e_1 \otimes \cdots \otimes e_n, f_1 \otimes \cdots \otimes f_n) = (e_1, f_1)_{H_1} \cdots (e_n, f_n)_{H_n}.$$

By definition the Hilbert tensor product $H_1 \otimes H_2 \otimes \cdots \otimes H_n$ is the completion of the algebraic tensor product with respect to this inner product.

Definition 10.3 Let $H_1, H_2, \ldots$ be an infinite sequence of Hilbert spaces with distinguished unit vectors $\xi_n \in H_n$ for all $n = 1, 2, \ldots$. Define an embedding of $H_1 \otimes H_2 \otimes \cdots \otimes H_n$ into $H_1 \otimes H_2 \otimes \cdots \otimes H_{n+1}$ by $\eta \mapsto \eta \otimes \xi_{n+1}$. Then the union $\bigcup_{n \geq 1} H_1 \otimes H_2 \otimes \cdots \otimes H_n$ is a pre-Hilbert space, and we denote by $\bigotimes_{n=1}^{\infty} H_n$ its completion. The latter Hilbert space is called the *infinite tensor product of Hilbert spaces H_n with stabilizing system* $\{\xi_n\}$. Let us emphasize that the construction substantially depends on the choice of the stabilizing system of vectors ξ_n.

By the very construction, a finite tensor product space $H_1 \otimes H_2 \otimes \cdots \otimes H_n$ is canonically embedded into the infinite tensor product space $\bigotimes_{k=1}^{\infty} H_k$. The image of a vector $\eta \in H_1 \otimes H_2 \otimes \cdots \otimes H_n$ under that embedding is denoted as $\eta \otimes \xi_{n+1} \otimes \xi_{n+2} \otimes \cdots$. In particular, in the infinite tensor product space, there is a distinguished vector $\xi = \xi_1 \otimes \xi_2 \otimes \cdots$.

Example 10.4 Infinite tensor products of Hilbert spaces naturally arise in the following situation. Let (X_1, P_1), (X_2, P_2), ... be a sequence of probability spaces (that is, X_n is a measurable space equipped with a probability measure P_n). Set $H_n = L^2(X_n, P_n)$. The finite tensor product $H_1 \otimes \cdots \otimes H_n$ can be identified with the Hilbert space $L^2(X_1 \times \cdots \times X_n, P_1 \otimes \cdots \otimes P_n)$. Now consider the infinite product space $X = X_1 \times X_2 \times \cdots$ with the product measure $P = P_1 \otimes P_2 \otimes \cdots$. Then the Hilbert space $H = L^2(X, P)$ can be identified, in a natural way, with the infinite tensor product of the Hilbert spaces $H_n = L^2(X_n, P_n)$ with the stabilizing system $\{\xi_n \equiv 1\}$ (constant functions on X_ns). The image of $H_1 \otimes \cdots \otimes H_n$ in H coincides with the subspace of the cylinder functions $f(x_1, x_2, \ldots)$ depending on the first n arguments only. The distinguished vector $\xi = \bigotimes_{n=1}^{\infty} \xi_n$ is the constant function $\xi \equiv 1$ on the infinite product space.

After these preliminaries we turn to the construction of the irreducible spherical representations. Let us start with the case when $\beta = \underline{0}$ and $\sum \alpha_i = 1$.

Denote by I the set of indices $i = 1, 2, \dots$ for which α_i is nonzero; thus, I may be finite or countable. Fix a Hilbert space with an orthonormal basis $\{e_i\}$ indexed by the set I, and let $\bar{E}$ be the dual Hilbert space with the dual basis $\{\bar{e}_i\}$.[1] For all $n = 1, 2, \dots$ we take as H_n the space $E \otimes \bar{E}$ with the distinguished unit vector

$$\xi_n = v = \sum_{i \in I} \sqrt{\alpha_i} \cdot e_i \otimes \bar{e}_i.$$

Note that $\|v\| = 1$ because of the assumption $\sum \alpha_i = 1$.

Define a representation T of $S(\infty) \times S(\infty)$ in $\bigotimes_{n=1}^{\infty} H_n$ by letting the first factor $S(\infty) \times \{e\}$ permute the copies of E, while the second factor $\{e\} \times S(\infty)$ permutes the copies of $\bar{E}$. In terms of basis vectors: if n is so large that $(\sigma, \tau) \in S(n) \times S(n)$ then for any basis vector $\eta \in \bigotimes_{k=n+1}^{\infty} H_k$,

$$\begin{aligned} T(\sigma^{-1}, \tau^{-1})\big((e_{i_1} \otimes \bar{e}_{j_1}) \otimes \cdots \otimes (e_{i_n} \otimes \bar{e}_{j_n}) \otimes \eta\big) \\ = \big((e_{i_{\sigma(1)}} \otimes \bar{e}_{j_{\tau(1)}}) \otimes \cdots \otimes (e_{i_{\sigma(n)}} \otimes \bar{e}_{j_{\tau(n)}}) \otimes \eta\big) \end{aligned}$$

(we inverted σ and τ in the left-hand side to avoid triple indices in the right-hand side). Clearly, the distinguished vector

$$\xi := \bigotimes_{n=1}^{\infty} v = \Big(\sum_{i \in I} \sqrt{\alpha_i} \cdot e_i \otimes \bar{e}_i\Big) \otimes \Big(\sum_{i \in I} \sqrt{\alpha_i} \cdot e_i \otimes \bar{e}_i\Big) \otimes \cdots$$

is invariant with respect to the action of the diagonal subgroup $\operatorname{diag} S(\infty)$.

Proposition 10.5 *Let $(\alpha, \underline{0}) \in \Omega$ be a point with $\sum \alpha_i = 1$ and T be the unitary representation of the group $S(\infty) \times S(\infty)$ constructed above. The subrepresentation of T realized in the closed cyclic span of the distinguished vector ξ is equivalent to the irreducible spherical representation of $T^{\alpha,\underline{0}}$ with parameters $(\alpha, \underline{0})$.*

Proof We have to prove that the matrix element $(T(\sigma, e)\xi, \xi)$ coincides with $\chi_\rho^{\alpha,\underline{0}}$, where ρ stands for the cycle structure of $\sigma \in S(\infty)$; that is, ρ is a finite sequence of numbers $\rho = (\rho_1 \geq \rho_2 \geq \cdots \geq 2)$ such that σ is the product

[1] One may think that $\bar{E}$ is obtained from E by replacing the complex structure by the conjugate one.

of disjoint cycles of lengths $\rho_1, \rho_2, \ldots$. Since $\sum_i \alpha_i = 1$ by assumption, we have to prove that

$$(T(\sigma, e)\xi, \xi) = \prod_{k\geq 1} \sum_{i\in I} \alpha_i^{\rho_k}.$$

Without loss of generality we may assume that σ is contained in $S(n)$ with $n = \rho_1+\rho_2+\cdots$. Replace σ by σ^{-1}, which does not affect the matrix element. We have

$$T(\sigma^{-1}, e)\xi$$
$$= \sum_{i_1,\ldots,i_n\in I} \sqrt{\alpha_{i_1}\cdots\alpha_{i_n}}\,(e_{i_{\sigma(1)}} \otimes \bar{e}_{i_1}) \otimes \cdots \otimes (e_{i_{\sigma(n)}} \otimes \bar{e}_{i_n}) \otimes v \otimes v \otimes \cdots,$$

$$\xi = \sum_{i_1,\ldots,i_n\in I} \sqrt{\alpha_{i_1}\cdots\alpha_{i_n}}\,(e_{i_1} \otimes \bar{e}_{i_1}) \otimes \cdots \otimes (e_{i_n} \otimes \bar{e}_{i_n}) \otimes v \otimes v \otimes \cdots,$$

whence

$$\begin{aligned}(T(\sigma^{-1}, e)\xi, \xi) &= \sum_{i_1,\ldots,i_n\in I} \alpha_{i_1}\cdots\alpha_{i_n}\,(e_{i_{\sigma(1)}}, e_{i_1})\cdots(e_{i_{\sigma(n)}}, e_{i_n}) \\ &= \sum_{i_1,\ldots,i_n\in I} \alpha_{i_1}\cdots\alpha_{i_n}\,\delta_{i_{\sigma(1)},i_1}\ldots\delta_{i_{\sigma(n)},i_n},\end{aligned}$$

which gives the desired result. □

In order to include nonzero $\beta = (\beta_1 \geq \beta_2 \geq \cdots)$ into this construction, we need to add orthonormal basis vectors f_j, corresponding to nonzero β_js, to the Hilbert space E. Of course, the dual vectors $\{\bar{f}_j\}$ are added to $\bar{E}$. The corresponding index set will be denoted as J.

We are about to define the action of $S(\infty) \times S(\infty)$ in $\bigotimes_{n=1}^{\infty} H_n$, where, as before, $H_n = E \otimes \bar{E}$ for all n, and the distinguished vectors $\xi_n \in H_n$ are all equal to

$$v = \sum_{i\in I} \sqrt{\alpha_i} \cdot e_i \otimes \bar{e}_i + \sum_{j\in J} \sqrt{\beta_j} \cdot f_j \otimes \bar{f}_j.$$

In contrast with the case considered above, this action involves a supplementary factor – a "cocycle" taking values ± 1.

To define this "cocycle" we introduce a $\mathbb{Z}_2$-grading in E and $\bar{E}$:

$$\begin{aligned} E &= E^{(0)} \oplus E^{(1)}, \qquad E^{(0)} = \mathrm{Span}\{e_i\}_{i\in I}, \quad E^{(1)} = \mathrm{Span}\{f_j\}_{j\in J} \\ \bar{E} &= \bar{E}^{(0)} \oplus \bar{E}^{(1)}, \qquad \bar{E}^{(0)} = \mathrm{Span}\{\bar{e}_i\}_{i\in I}, \quad \bar{E}^{(1)} = \mathrm{Span}\{\bar{f}_j\}_{j\in J} \end{aligned}$$

where Span denotes the closed linear span. The vectors from $E^{(0)}$ or $\bar{E}^{(0)}$ are called *even*, and the vectors from $E^{(1)}$ or $\bar{E}^{(1)}$ are called *odd*.

To define the symmetric group action on tensor products of $\mathbb{Z}_2$-graded vector spaces we use the general *sign rule*: for each pair of odd vectors that pass through each other, the result is multiplied by (-1).

For example, if V_1, V_2, and V_3 are three $\mathbb{Z}_2$-graded vector spaces then under the canonical isomorphism

$$V_1 \otimes V_2 \otimes V_3 \to V_2 \otimes V_3 \otimes V_1$$

we have

$$v_1 \otimes v_2 \otimes v_3 \to (-1)^{p(v_1)p(v_2)+p(v_1)p(v_3)}\, v_2 \otimes v_3 \otimes v_1$$

where $v_1 \in V_1$, $v_2 \in V_2$, and $v_3 \in V_3$ are assumed to be homogeneous vectors and $p(\cdot)$ stands for the *parity function* taking value 0 for even vectors and value 1 for odd vectors.

More generally, if $V_1, \dots, V_k$ are $\mathbb{Z}_2$-graded vector spaces and s is a permutation of $1, \dots, k$ then under the canonical isomorphism

$$V_1 \otimes \cdots \otimes V_k \to V_{s^{-1}(1)} \otimes \cdots V_{s^{-1}(k)}$$

we have for homogeneous vectors $v_i \in V_i$

$$v_1 \otimes \cdots \otimes v_k \to (-1)^m\, v_{s^{-1}(1)} \otimes \cdots \otimes v_{s^{-1}(k)}$$

where the integer m is computed as follows: write down the indices $i_1 < \cdots < i_l$ corresponding to odd vectors and then take the number of inversions in the subsequence $s^{-1}(i_1), \dots, s^{-1}(i_l)$.

Now, we apply the sign rule to define the action of the group $S(n) \times S(n)$ on the $2n$-fold tensor product

$$(E \otimes \bar{E})^{\otimes n} = E \otimes \bar{E} \otimes \cdots \otimes E \otimes \bar{E}.$$

The resulting action is consistent with the embedding $(E \otimes \bar{E})^{\otimes n} \to (E \otimes \bar{E})^{\otimes(n+1)}$ given by tensoring with v.[2] This makes it possible to define, as before, a unitary representation T of the group $S(\infty) \times S(\infty)$ in the infinite product space. Note that, as before, the vector $\xi = v \otimes v \otimes v \otimes \cdots$ is $\operatorname{diag} S(\infty)$-invariant.

The following claim is a generalization of Proposition 10.5:

Proposition 10.6 *Let $(\alpha, \beta) \in \Omega$ be a point with $\sum \alpha_i + \sum \beta_j = 1$ and T be the unitary representation of the group $S(\infty) \times S(\infty)$ constructed above. The subrepresentation of T realized in the closed cyclic span of the distinguished vector ξ is equivalent to the irreducible spherical representation $T^{\alpha,\beta}$ with parameters (α, β).*

[2] The crucial point here is that v is an even vector with respect to the natural $\mathbb{Z}_2$-grading of $H = E \otimes \bar{E}$.

Proof The argument is similar to that in the proof of Proposition 10.5. To simplify the control of signs while computing the matrix element it is convenient to rearrange the factors in $(E \otimes \bar{E})^{\otimes n}$ and pass from this space to $E^{\otimes n} \otimes \bar{E}^{\otimes n}$. Due to the sign rule, instead of the ordinary power sums $\sum \alpha_i^r$ we will get $\sum \alpha_i^r + (-1)^{r-1} \sum \beta_j^r$: specifically, the sign $(-1)^{r-1}$ appears as a result of a length r cycle acting on the tensor product of r odd vectors. □

Finally, let us discuss the case when $0 < \sum \alpha_i + \sum \beta_j < 1$. The simplest example is that of

$$\alpha = (p, 0, 0, \ldots), \quad \beta = \underline{0} = (0, 0, \ldots), \qquad 0 < p < 1.$$

Denote the corresponding irreducible spherical representation as $T^{[p]}$.

Proposition 10.7 *Let $(\alpha, \beta) \in \Omega$ be an arbitrary point and $0 < p < 1$. Consider the tensor product representation $T = T^{[p]} \otimes T^{\alpha,\beta}$ and denote by ξ its vector obtained as the tensor product of the spherical vectors of the factors. The subrepresentation of T realized in the closed cyclic span of ξ is equivalent to the irreducible spherical representation $T^{\alpha',\beta'}$ with parameters*

$$\alpha' = p\alpha = (p\alpha_1, p\alpha_2, \ldots), \quad \beta' = p\beta = (p\beta_1, p\beta_2, \ldots).$$

Proof Clearly, ξ is an invariant vector with respect to diag$S(\infty)$. Its matrix element equals the product of the spherical functions of the representations $T^{[p]}$ and $T^{\alpha,\beta}$. This immediately implies the desired result. (cf. Exercise 4.4.) □

Thus, to construct an arbitrary irreducible spherical representation it remains to exhibit a realization of the representations $T^{[p]}$. Such a realization is proposed in Exercise 10.2.

10.1 Exercises

Exercise 10.1 Let H be the infinite tensor product of Hilbert spaces $H_1, H_2, \ldots$ with a stabilizing sequence of unit vectors $\{\xi_i\}$. Denote by $\mathbb{N}$ the set of positive integers. Given a collection of vectors $\eta_1 \in H_1, \eta_2 \in H_2, \ldots$ and a finite subset $S \subset \mathbb{N}$, we define a vector $\eta_S \in H$ in the following way:

$$\eta_S := \widetilde{\eta}_1 \otimes \widetilde{\eta}_2 \otimes \cdots,$$

where

$$\widetilde{\eta}_i := \begin{cases} \eta_i, & i \in S \\ \xi_i, & i \in \mathbb{N} \setminus S. \end{cases}$$

Let $\{S\}$ be the directed set of finite subsets of $\mathbb{N}$ ordered by inclusion. We define the infinite tensor product vector $\eta = \bigotimes_{i\in\mathbb{N}} \eta_i \in H$ as a limit taken over $\{S\}$,

$$\eta = \bigotimes_{i\in\mathbb{N}} \eta_i := \lim_{\{S\}} \eta_S, \tag{10.1}$$

provided this limit exists.

(a) Show that η exists if and only if

$$\sum_{k=1}^{\infty} \big| \|\eta_k\| - 1 \big| < \infty \quad \text{and} \quad \sum_{k=1}^{\infty} \big| (\eta_k, \xi_k) - 1 \big| < \infty.$$

(b) Show that if two infinite products $\eta = \bigotimes_{i\in\mathbb{N}} \eta_i$ and $\eta' = \bigotimes_{i\in\mathbb{N}} \eta'_i$ exist, then the inner product (η, η') is given by the absolutely converging product $\prod_{k=1}^{\infty} (\eta_k, \eta'_k)$.

(c) Let H' be the infinite tensor product of the same spaces but built with a different stabilizing system $\{\xi'_i\}$. Let us say that $\{\xi'_i\}$ is *equivalent* to the initial stabilizing system $\{\xi_i\}$ if the collections $\{\eta_i\}$ for which the limit (10.1) exists are the same for H and for H'.

Show that $\{\xi_i\}$ and $\{\xi'_i\}$ are equivalent if and only if $\sum |(\xi_i, \xi'_i) - 1| < \infty$. Deduce from this that equivalence implies existence of a (unique) isometry $H \to H'$ preserving the set of infinite tensor products of the form (10.1). In this sense, the infinite tensor product of Hilbert spaces depends only on the equivalence class of the stabilizing system.

Exercise 10.2 Here we describe a realization of the representation $T^{[p]}$; see the end of Chapter 10. We need the values $p \in (0, 1)$ but below we assume, slightly more generally, that $0 \le p \le 1$.

A matrix with entries in the two-point set $\{0, 1\}$ is said to be *monomial* if it contains at most one 1 in each row and each column. Let $\mathfrak{X}$ denote the set of all monomial 0–1 matrices $\varepsilon = [\varepsilon_{ij}]$ of infinite size, containing finitely many 1s outside the diagonal. This set can be written as the union $\bigcup_{n\ge 0} \mathfrak{X}_n$ of ascending subsets $\mathfrak{X}_n$, where $\mathfrak{X}_n$ consists of the matrices $\varepsilon \in \mathfrak{X}$ such that $\varepsilon_{ij} = 0$ for all $i \ne j$ with $\max(i, j) > n$. In particular, $\mathfrak{X}_0 \subset \mathfrak{X}$ is the subset formed by the diagonal 0–1 matrices.

The set $\mathfrak{X}_0$ can be identified with $\{0, 1\}^\infty$, the set of all infinite binary sequences. Let us equip it with the Bernoulli probability measure ν_p^0 such that, for any i, the value of ε_{ii} equals 0 or 1 with probabilities p and $1 - p$, respectively.

For any n, we extend this measure to a measure ν_p^n on $\mathfrak{X}_n$ as follows. Denote by $\mathfrak{X}^{(n)}$ the finite set of monomial 0–1 matrices of size $n \times n$. We equip it with the measure $\nu_p^{(n)}$ such that the weight $\nu_p^{(n)}(\eta)$ of a matrix $\eta \in \mathfrak{X}^{(n)}$ equals

$(1-p)^m p^{n-m}$, where m stands for the number of 1s in η (if p equals 0 or 1 we have to define the meaning of the symbol 0^0: we agree that it equals 1).

There is a natural bijection between $\mathfrak{X}_n$ and the product space $\mathfrak{X}^{(n)} \times \{0, 1\}^\infty$:

$$\mathfrak{X}_n \ni \varepsilon \leftrightarrow \left([\varepsilon_{ij}]_{i,j=1}^n, \quad \{\varepsilon_{kk}\}_{k=n+1}^\infty\right) \in \mathfrak{X}^{(n)} \times \{0, 1\}^\infty.$$

By definition, ν_p^n is the image, under this bijection, of the product of the measure $\nu_p^{(n)}$ on $\mathfrak{X}^{(n)}$ with the same Bernoulli measure on $\{0, 1\}^\infty$ as before.

(a) Check that the measures ν_p^n are pairwise consistent in the sense that the restriction of ν_p^{n+1} to the subset $\mathfrak{X}_n \subset \mathfrak{X}_{n+1}$ coincides with the measure ν_p^n. Deduce from this that there exists an (infinite) measure ν_p on the set $\mathfrak{X}$ such that for any $n = 0, 1, 2, \dots$, the restriction of ν_p to $\mathfrak{X}_n$ coincides with ν_p^n. In particular, the restriction to $\mathfrak{X}_0$ is the Bernoulli measure.

(b) Show that ν_1 is the delta measure concentrated at the null matrix (all entries equal 0).

(c) Show that ν_0 is the counting measure on the subset $S(\infty) \subset \mathfrak{X}$. Here we employ the natural embedding of $S(\infty)$ into $\mathfrak{X}$ which assigns to a permutation $\sigma \in S(\infty)$ the matrix ε_{ij} such that ε_{ij} takes value 1 precisely when $\sigma(j) = i$.

(d) Observe that the group $G = S(\infty) \times S(\infty)$ acts on the set $\mathfrak{X}$ by permutations of rows and columns. Prove that ν_p is an invariant measure for this action. Use this to define a unitary representation T of this group in the Hilbert space $L^2(\mathfrak{X}, \nu_p)$.

(e) Observe that for $p = 1$ this representation is the trivial one while for $p = 0$ it is the biregular representation.

(f) Assume $0 < p < 1$. Let ξ_0 be the characteristic function of the subset $\mathfrak{X}_0$. Show that ξ_0 is a K-invariant vector and that the corresponding matrix coefficient is the spherical function of the representation $T^{[p]}$. Thus, the cyclic span of ξ_0 is equivalent to $T^{[p]}$.

10.2 Notes

Our exposition in this section follows Olshanski's paper [85]. We have already pointed out in the Introduction that our realization of the irreducible spherical representations T^ω was obtained by a modification of a construction due to Vershik and Kerov [121]. The same construction with infinite tensor products was sketched in Wassermann's dissertation [132] (it is not cited in [85] because Olshanski received a copy of [132] from Robert Boyer in 1991, after the paper [85] was published).

As explained in [85], one can define a reasonable category of unitary representations of the group $S(\infty) \times S(\infty)$, which are well behaved with respect to restriction to the subgroup diag$S(\infty)$; they are called *admissible representations of the pair* $(S(\infty) \times S(\infty), \text{diag} S(\infty))$. All spherical representations are admissible. Irreducible admissible representations admit a complete classification; see Olshanski [85] and Okounkov [78], [79]. A refinement of the infinite tensor product construction provides an explicit realization of the irreducible admissible representations (see [85]), and so does Okounkov's construction (see [78], [79]).

The notion of infinite tensor product of Hilbert spaces with a distinguished stabilizing sequence of vectors is due to von Neumann [77]. We used an obvious extension of the conventional definition to $\mathbb{Z}_2$-graded (= super) Hilbert spaces. For the basics of linear superalgebras see, e.g., Manin [73, ch. 3].

The construction of Exercise 10.2 is due to Vershik and Kerov.

11

Generalized Regular Representations T_z

In Proposition 10.1 we saw that the biregular representation of $S(\infty)\times S(\infty)$ in $\ell^2(S(\infty))$ is irreducible. This is in sharp contrast with the situation in representation theory of finite groups, where decomposing the biregular representation is essentially equivalent to finding all irreducible representations of the corresponding group. The objective of this chapter is to construct a (unitary spherical) deformation of the biregular representation of $(S(\infty)\times S(\infty), \operatorname{diag} S(\infty))$ for which the above-mentioned problem of harmonic analysis is meaningful.

As before, we fix a sequence of embeddings

$$S(1) \subset S(2) \subset \cdots \subset S(n) \subset \cdots \subset S(\infty),$$

where $S(n)$ is viewed as the subgroup of $S(n+1)$ that permutes only the first n symbols.

For any $n = 1, 2, \ldots$ define a map $p_{n,n+1} : S(n+1) \to S(n)$ by

$$p_{n,n+1}(\sigma)(i) = \begin{cases} \sigma(i), & \sigma(i) \le n, \\ \sigma(n+1), & \sigma(i) = n+1, \end{cases}$$

for all $i = 1, \ldots, n$. In other words, $p_{n,n+1}$ acts by striking out $n+1$ from the corresponding cycle $(\cdots \to n+1 \to \cdots)$ of the permutation σ.

The maps $p_{n,n+1}$ are called *canonical projections.*

Proposition 11.1 *The canonical projection $p_{n,n+1}$ commutes with the action of $S(n)\times S(n)$ on $S(n+1)$ and $S(n)$ by left and right shifts. For $n \ge 4$, $p_{n,n+1}$ is the only map $S(n+1) \to S(n)$ with this property.*

Proof It is convenient to represent permutations $\sigma \in S(n+1)$ as bipartite graphs with vertices $\{1, \ldots, n+1\}$ and $\{1', \ldots, (n+1)'\}$ and edges joining i and $\sigma(i)'$. Then the application of $p_{n,n+1}$ is equivalent to adding an extra edge joining $(n+1)$ and $(n+1)'$ and contracting the resulting sequence of three edges (if $\sigma \notin S(n)$) or two edges (if $\sigma \in S(n)$) to a single one. Multiplying σ on the left or on the right by an element of $S(n)$ is equivalent to adding edges that connect $\{1, \dots, n\}$ and $\{1', \dots, n'\}$ to some new vertices and contracting the intermediate edges. One easily sees that this procedure commutes with adding the extra edge joining $(n+1)$ and $(n+1)'$.

Let us prove the uniqueness. Assume $p : S(n+1) \to S(n)$ is a map with the desired property. Then $\sigma^{-1} p(e) \sigma = p(e)$ for any $\sigma \in S(n)$. Since for $n \geq 3$, e is the only central element of $S(n)$, we have $p(e) = e = p_{n,n+1}(e)$. Similarly, $\tau^{-1} p((n, n+1)) \tau = p((n, n+1))$ for any $\tau \in S(n-1)$. If $n \geq 4$ then e is the only element of $S(n)$ commuting with $S(n-1)$. Thus, $p((n, n+1)) = e = p_{n,n+1}((n, n+1))$. But the group $S(n+1)$ is made of just two double $S(n) \times S(n)$-cosets, and e and $(n, n+1)$ lie in different cosets. □

Denote by $\mathfrak{S}$ the projective limit of $S(n)$s with respect to the system of canonical projections: $\mathfrak{S} = \varprojlim S(n)$. In other words, the elements of $\mathfrak{S}$ are the sequences $\{x_n\}_{n=1}^{\infty}$ with $x_n \in S(n)$ such that $p_{n,n+1}(x_{n+1}) = x_n$ for any $n \geq 1$. The space $\mathfrak{S}$ is equipped with the projective limit topology: the base of the topology at the point $\{x_n\} \in \mathfrak{S}$ contains cylinder sets in $\mathfrak{S}$ that coincide with $\{x_n\}$ on the first few coordinates.

One of the exercises at the end of this chapter is to show that $\mathfrak{S}$ is compact; this also follows from Proposition 11.2, below.

The group $S(\infty)$ is embedded in $\mathfrak{S}$ as an everywhere dense discrete subset of stabilizing sequences. Thus, $\mathfrak{S}$ can be viewed as a compactification of $S(\infty)$. Of course, this is only a set compactification, not a group compactification: $\mathfrak{S}$ is no longer a group. The elements of $\mathfrak{S}$ are called *virtual permutations* of $\{1, 2, \dots\}$.

Proposition 11.2 *There exists a natural homeomorphism between the space* $\mathfrak{S}$ *and the infinite product space*

$$I = I_1 \times I_2 \times \cdots, \qquad I_n = \{0, 1, \dots, n-1\}.$$

Proof Take a sequence $\{x_n\} \in \mathfrak{S}$ and let $\{i_n\} \in I$ be its image in I to be described. Then the coordinate i_{n+1} serves to specify the choice of x_{n+1} among the $n+1$ elements in $p_{n,n+1}^{-1}(x_n) \subset S(n+1)$. Specifically, $i_{n+1} = 0$ means that x_{n+1} coincides with x_n (which is equivalent to saying that $n+1$ forms in x_{n+1} a new trivial cycle), and $i_{n+1} = j \in \{1, \dots, n\}$ means that in order to get x_{n+1} from x_n one inserts $n+1$ into the cycle of x_n containing the point j and immediately before it. One readily checks that this is a homeomorphism. □

Clearly, under this homeomorphism the canonical projection $p_{n,n+1}$ simply turns into the operation of ignoring the last coordinate of an element in $I_1 \times \cdots \times I_{n+1}$.

As before, we are going to use the notation

$$G(n) = S(n) \times S(n), \quad K(n) = \mathrm{diag} S(n), \quad G = S(\infty) \times S(\infty), \quad K = \mathrm{diag} S(\infty).$$

The group G acts on $\mathfrak{S}$ by homeomorphisms generated by left and right shifts of G on itself: for any $\{x_n\} \in \mathfrak{S}$ and $g = (g_1, g_2) \in G$ we pick N so large that $g \in G(N)$ and for $n \geq N$ define

$$(x \cdot g)_n = g_2^{-1} x_n g_1.$$

This automatically defines all the coordinates of $x \cdot g$ with smaller indices.

In what follows we denote by $[\sigma]$ the number of cycles of a permutation $\sigma \in S(n)$. Let us also denote by $p_n : \mathfrak{S} \to S(n)$ the natural projection that maps the sequence $\{x_k\}_{k=1}^{\infty} \in \mathfrak{S}$ to $x_n \in S(n)$.

Proposition 11.3 *Let $x \in \mathfrak{S}$ and $g \in G$. The quantity $[p_n(x \cdot g)] - [p_n(x)]$ stabilizes for large n. Specifically, if n is so large that $g \in G(n)$, then $[p_n(x \cdot g)] - [p_n(x)]$ does not depend on n.*

Proof It suffices to consider $g = (\sigma, e)$ and $g = (e, \sigma)$ with $\sigma = (ij)$ (transposition of i and j). Assume $n \geq \max(i, j)$. If i and j are in the same cycle of $x_n \in S(n)$ then the multiplication of x_n by (ij) on the left or on the right splits this cycle into two; otherwise the two cycles containing i and j merge. In either case, $[x_n \cdot g] - [x_n] = \pm 1$, and the result does not depend on n. □

We set

$$c(x, g) = \text{stable value of } [p_n(x \cdot g)] - [p_n(x)], \qquad x \in \mathfrak{S}, \quad g \in G,$$

and call $c(x, g)$ the *fundamental cocycle*. It is immediate to see that it satisfies the additive 1-cocycle relation:

$$c(x, g_1 g_2) = c(x, g_1) + c(x \cdot g_1, g_2), \qquad x \in \mathfrak{S}, \ g_1, g_2 \in G. \tag{11.1}$$

Clearly, $c(\,\cdot\,, g) \equiv 0$ for any $g \in K$.

We use the fundamental cocycle below in Proposition 11.6.

Proposition 11.4 *There exists a unique G-invariant probability measure on $\mathfrak{S}$.*

This measure serves as an analog of the Haar measure; we denote it by μ_1.

Proof Let μ_1^n denote the uniform (Haar) probability measure on $S(n)$. Clearly, μ_1^n coincides with the pushforward of μ_1^{n+1} under the canonical projection $p_{n,n+1}$. Thus, by virtue of Theorem 7.9, the sequence $\{\mu_1^n\}_{n\geq 1}$ correctly defines a probability measure μ_1 on $\mathfrak{S}$, and one readily sees that it is G-invariant.

Conversely, let μ be a G-invariant probability measure on $\mathfrak{S}$. Its pushforward under the projection $p_n : \mathfrak{S} \to S(n)$ must coincide with μ_1^n, which is the only $G(n)$-invariant probability measure on $S(n)$. Hence, $\mu = \mu_1$. □

We now construct a one-parameter family $\{\mu_t\}_{t>0}$ of probability measures on $\mathfrak{S}$ that are *quasi-invariant* with respect to the G-action.

For $t > 0$ and $n = 1, 2, \ldots$, we define a measure μ_t^n on $S(n)$ by

$$\mu_t^n(\{\sigma\}) = \frac{t^{[\sigma]}}{t(t+1)\cdots(t+n-1)}, \qquad \sigma \in S(n).$$

Note that the encoding of Proposition 11.2 also determines a bijection $S(n) \leftrightarrow I_1 \times \cdots \times I_n$, for any $n = 1, 2, \ldots$.

Proposition 11.5 *Under the above bijection, μ_t^n turns into a product measure $\widetilde{\mu}_t^1 \otimes \cdots \otimes \widetilde{\mu}_t^n$, where $\widetilde{\mu}_t^m$ is the measure on $I_m = \{0, 1, \ldots, m-1\}$ defined by*

$$\widetilde{\mu}_t^m(\{k\}) = \begin{cases} \frac{1}{t+m-1}, & k = 1, \ldots, m-1, \\ \frac{t}{t+m-1}, & k = 0. \end{cases}$$

Proof Follows from the fact that the number of cycles of any $\sigma \in S(n)$ coincides with the number of 0's in its image in $I_1 \times \cdots \times I_n$. □

Proposition 11.5 implies that μ_t^n is a probability measure (since all $\widetilde{\mu}_t^m$ are), and that μ_t^ns are consistent with canonical projections. By virtue of Theorem 7.9, the sequence $\{\mu_t^n\}$ defines a probability measure on $\mathfrak{S}$ that we denote by μ_t and call the *Ewens measure*. Under the homeomorphism of Proposition 11.2, μ_t is simply the product measure $\widetilde{\mu}_t^1 \otimes \widetilde{\mu}_t^2 \otimes \cdots$ on I. One readily sees that μ_t is K-invariant. A few other interesting properties of the Ewens measures can be found in the exercises to this chapter.

Recall that two measures ν_1 and ν_2 defined on the same space are called *equivalent* if they have the same set of null subsets. Then, by virtue of the Radon–Nikodým theorem, $\nu_2 = f\nu_1$, where f is a function, called the *Radon–Nikodým derivative* and denoted as ν_2/ν_1. This function is defined uniquely within a null set (with respect to ν_1). A measure is called *quasi-invariant* with respect to a group action if its shifts by elements of the group remain in the same equivalence class of measures.

Proposition 11.6 *Each of the Ewens measures μ_t, $0 < t < \infty$, is quasi-invariant with respect to the G-action on $\mathfrak{S}$. More precisely, the Radon–Nikodým derivative is given by*

$$\frac{\mu_t(dx \cdot g)}{\mu_t(dx)} = t^{c(x,g)}, \qquad x \in \mathfrak{S}, \quad g \in G,$$

where $c(x, g)$ is the fundamental cocycle.

Proof It suffices to check that

$$\mu_t(V \cdot g) = \int_V t^{c(x,g)} \mu_t(dx)$$

for every Borel subset $V \subset \mathfrak{S}$ and any $g \in G$.

Fix $g \in G$ and choose m so large that $g \in G(m)$. For arbitrary $n \geq m$ and $y \in S(n)$, let $V_n(y) \subset \mathfrak{S}$ denote the set of elements $\{x_n\} \in \mathfrak{S}$ with $x_n = y$. This is a cylinder set, and any cylinder set is a disjoint union of sets of the form $V_n(y)$, $n \geq m$. Since Borel sets of $\mathfrak{S}$ are generated by cylinder sets, it suffices to check the above formula for $V = V_n(y)$.

Note that $V_n(y) \cdot g = V_n(y \cdot g)$ and $\mu_t(V_n(y)) = \mu_t^n(\{y\})$, hence

$$\mu_t(V_n(y) \cdot g) = \mu_t^n(\{y \cdot g\}).$$

On the other hand, the definition of μ_t^n implies

$$\mu_t^n(\{y \cdot g\}) = t^{[y \cdot g]-[y]} \mu_t^n(\{y\}),$$

and this is exactly what we need, because $c(x, g) = [y \cdot g] - [y]$ for any $x \in V_n(y)$. □

We are finally ready to introduce new representations of (G, K). In what follows we fix $t > 0$ and $z \in \mathbb{C}$ such that $|z|^2 = t$. The formula of Proposition 11.6 then can be written as

$$\frac{\mu_t(dx \cdot g)}{\mu_t(dx)} = \left|z^{c(x,g)}\right|^2.$$

The fact that $c(x, g)$ satisfies the additive cocycle relation, see (11.1) above, implies that the formula

$$(T_z(g)f)(x) = f(x \cdot g)\, z^{c(x,g)}, \qquad g \in G, \quad x \in \mathfrak{S},$$

correctly defines an action of the group G on functions on the space $\mathfrak{S}$. Moreover, Proposition 11.6 implies that the operators $T_z(g)$ are unitary operators in the Hilbert space $L^2(\mathfrak{S}, \mu_t)$, $t = |z|^2$. Thus, we get a unitary representation T_z in $L^2(\mathfrak{S}, \mu_t)$. We call it the *generalized regular representation* of G with parameter z. One explanation for the name comes from the fact that T_z can be realized as an inductive limit of the regular representations of $G(n)$ as follows.

For every $n = 1, 2, \ldots$ denote by H^n the finite-dimensional space $L^2(S(n), \mu_1^n)$, and denote by Reg^n the biregular representation of the group $G(n) = S(n) \times S(n)$ in this space:

$$(\mathrm{Reg}^n(g)\, f)(x) = f(g_2^{-1}\, x\, g_1), \quad g = (g_1, g_2) \in G(n), \quad x \in S(n), \quad f \in H^n.$$

Define the operators $L_z^n \colon H^n \to H^{n+1}$ as follows: for $f \in H^n$ and $x \in S(n+1)$,

$$(L_z^n f)(x) = \begin{cases} z\sqrt{\frac{n+1}{t+n}}\, f(x), & x \in S(n) \subset S(n+1), \\ \sqrt{\frac{n+1}{t+n}}\, f(p_{n,n+1}(x)), & x \in S(n+1) \setminus S(n). \end{cases}$$

Proposition 11.7 *For any $z \in \mathbb{C}^*$ the operator L_z^n provides an isometric embedding $H^n \to H^{n+1}$ which intertwines the representations* Reg^n *and* $\mathrm{Reg}^{n+1}\big|_{G(n)}$ *of the group $G(n)$. The generalized regular representation T_z is equivalent to the inductive limit of the representations* Reg^n *with respect to these embeddings.*

Proof For every $n = 1, 2, \ldots$ the subspace $\mathrm{Cyl}^n \subset \mathscr{H} = L^2(\mathfrak{S}, \mu_t)$ of cylinder functions of level n (that is, functions $f(x)$ depending on $p_n(x)$ only) is invariant with respect to the operators $T_z(g)$, $g \in G(n)$. Indeed, for all $g \in G(n)$ the function $x \mapsto c(x, g)$ is a cylinder function of level n. Thus, we can identify the Hilbert spaces $\mathrm{Cyl}^n \subset \mathscr{H}$ and $L^2(S(n), \mu_t^n)$. The operators $T_z(g)\big|_{\mathrm{Cyl}^n}$ take the form

$$(T_z(g)\, f)(x) = f(x{\cdot}g)\, z^{[x\cdot g]-[x]}, \quad g \in G(n), \quad x \in S(n), \quad f \in L^2(S(n), \mu_t^n).$$

Define a function F_z^n on the group $S(n)$ by the formula

$$F_z^n(x) = \left(\frac{n!}{t(t+1)\ldots(t+n-1)}\right)^{\frac{1}{2}} z^{[x]}, \qquad x \in S(n).$$

Observe that

$$(T_z(g)\, f)(x) = f(x \cdot g)\, \frac{F_z^n(x \cdot g)}{F_z^n(x)}.$$

Further, the function $|F_z^n(x)|^2$ coincides with the density of the measure μ_t^n with respect to the uniform measure μ_1^n. It follows that the operator of multiplication by the function F_z^n defines an isometry

$$\mathrm{Cyl}^n = L^2(S(n), \mu_t^n) \longrightarrow L^2(S(n), \mu_1^n) = H^n$$

which commutes with the action of the group $G(n)$.

Consider now the commutative diagram

$$\begin{array}{ccc} L^2(S(n),\, \mu_t^n) = \mathrm{Cyl}^n & \longrightarrow & \mathrm{Cyl}^{n+1} = L^2(S(n+1),\, \mu_t^{n+1}) \\ \downarrow & & \downarrow \\ H^n & \xrightarrow{\widetilde{L}_z^n} & H^{n+1} \end{array}$$

where the top arrow denotes the natural embedding (lifting of functions via the projection $p_{n,n+1}$), the vertical arrows correspond to multiplication by F_z^n and F_z^{n+1}, respectively, and the bottom arrow $\widetilde{L}_z^n$ is defined by the commutativity requirement. Hence, for $f \in H^n$ and $x \in S(n+1)$ we obtain

$$\begin{aligned} (\widetilde{L}_z^n f)(x) &= F_z^{n+1}(x) \left(F_z^n(p_{n,n+1}(x)) \right)^{-1} f(p_{n,n+1}(x)) \\ &= \sqrt{\frac{n+1}{t+n}}\, z^{[x]-[p_{n,n+1}(x)]}\, f(p_{n,n+1}(x)). \end{aligned}$$

This implies that the operators $\widetilde{L}_z^n$ are isometric embeddings. Next, this also shows that the representation T_z is equivalent to the inductive limit of the representations Reg^n corresponding to the embeddings $\widetilde{L}_z^n$.

Finally, we observe that the above expression for $\widetilde{L}_z^n$ coincides with that for L_z^n. Indeed, this follows from the fact that

$$[x] = \begin{cases} [p_{n,n+1}(x)] + 1, & \text{if } x \in S(n) \subset S(n+1), \\ [p_{n,n+1}(x)], & \text{if } x \in S(n+1) \setminus S(n). \end{cases}$$

□

Note that the above construction can be extended to the limit points $z = 0$ and $z = \infty$; in particular, at $z = \infty$ one obtains the biregular representation of $S(\infty) \times S(\infty)$ in $\ell^2(S(\infty))$ (see Exercise 11.7). This is a justification of the term "generalized regular representations" that we use for T_z.

Observe that the representation T_z comes with a distinguished vector $\xi_0 \in L^2(\mathfrak{S}, \mu_t)$, which is the function identically equal to 1. Clearly, ξ_0 is K-invariant and $\|\xi_0\| = 1$.

The closed cyclic span of this vector under the action of G is the space of a spherical representation of the Gelfand pair (G, K). According to the general formalism, the corresponding spherical function gives rise to a character χ_z of $S(\infty)$ (Proposition 8.19) that in its turn leads to a coherent system of distributions $\{M_z^{(n)}\}_{n\geq 0}$ on $\{\mathbb{Y}_n\}_{n\geq 0}$; see Definition 3.4. Let us compute these distributions.

Recall the notation $c(\square) = j - i$ for a box $\square = (i, j)$ of a Young diagram.

Proposition 11.8 *Let $z \in \mathbb{C}^*$ and $t = |z|^2$. For any $\lambda \in \mathbb{Y}_n$,*

$$M_z^{(n)}(\lambda) = \frac{\prod_{\square\in\lambda} |z + c(\square)|^2}{t(t+1)\dots(t+n-1)} \frac{\dim^2 \lambda}{n!}, \tag{11.2}$$

where the product is taken over all boxes of the Young diagram λ.

Proof We use the realization of T_z as the inductive limit of representations Reg^n as described in Proposition 11.7. The distinguished vector ξ_0 belongs to H^1, hence to all of H^n. As an element of H^n it coincides with the function F_z^n. Therefore, for $\sigma \in S(n)$

$$\chi_z|_{S(n)}(\sigma) = \left(\mathrm{Reg}^n(\sigma, e)\, F_z^n, F_z^n\right) = \frac{1}{n!} \sum_{\tau\in S(n)} F_z^n(\tau\sigma)\, \overline{F_z^n(\tau)}$$

or, in other words, $\chi_z|_{S(n)} = (F_z^n)^* * F_z^n$ (recall that $f^*(\tau) = \overline{f(\tau^{-1})}$ denotes the involution in the group algebra).

Note that F_z^n is a central function on $S(n)$; hence it can be written as a linear combination of irreducible characters χ^λ (here we use the term "character" in the conventional sense),

$$F_z^n = \sum_{\lambda\in\mathbb{Y}_n} a_\lambda\, \chi^\lambda,$$

with some coefficients $a_\lambda \in \mathbb{C}$. The orthogonality relations for irreducible characters (see Proposition 1.5) immediately imply that $M_z^{(n)}(\lambda) = |a_\lambda|^2$. It remains to find a_λs. This is done in the following statement, which implies the proposition. □

Lemma 11.9 *For any $n = 1, 2, \dots$ and $z \in \mathbb{C}^*$, the expansion of the central function $x \mapsto z^{[x]}$ on the group $S(n)$ in the basis of the irreducible characters $\{\chi^\lambda\}_{\lambda\in\mathbb{Y}_n}$ has the form*

$$z^{[x]} = \sum_{\lambda\in\mathbb{Y}_n} \prod_{\square\in\lambda} (z + c(\square)) \cdot \frac{\dim\lambda}{n!} \chi^\lambda(x) = \sum_{\lambda\in\mathbb{Y}_n} \prod_{\square\in\lambda} \frac{z + c(\square)}{h(\square)} \chi^\lambda(x).$$

Here $h(\square)$ is the length of the hook associated to $\square$; see Chapter 1.

Proof The argument is based on the characteristic map described at the end of Chapter 2. Applying this map we see that our claim is equivalent to the formula

$$\mathrm{ch}(z^{[\cdot]}) = \sum_{\lambda\in\mathbb{Y}_n} \prod_{\square\in\lambda} \frac{z + c(\square)}{h(\square)} \cdot s_\lambda,$$

where s_λ are the Schur functions.

Denote by $y_1, y_2, \ldots$ a sequence of formal variables of symmetric functions, and let u be an additional formal variable. Recall (see Exercise 2.14) the formula for the number of elements in the conjugacy class C_ρ of $S(n)$:

$$|C_\rho| = \frac{n!}{1^{m_1} m_1! 2^{m_2} m_2! \cdots},$$

where $\rho = 1^{m_1} 2^{m_2} \ldots$ (i.e., ρ has m_1 parts equal to 1, m_2 parts equal to 2 etc.). Using it we obtain

$$1 + \sum_{n \geq 1} \operatorname{ch}(z^{[\cdot]})\, u^n = \sum_{\rho = (1^{m_1} 2^{m_2} \ldots)} \frac{z^{m_1 + m_2 + \ldots}\, u^{1m_1 + 2m_2 + \ldots}}{1^{m_1}\, 2^{m_2} \ldots m_1!\, m_2! \ldots} p_1^{m_1} p_2^{m_2} \ldots =$$

$$= \prod_{n=1}^{\infty} \sum_{k=0}^{\infty} \frac{z^k\, p^k\, u^{nk}}{n^k k!} = \exp z \left(\frac{u p_1}{1} + \frac{u^2 p_2}{2} + \ldots \right) =$$

$$= \exp z \sum_{i=1}^{\infty} \left(\frac{u y_i}{1} + \frac{(u y_i)^2}{2} + \ldots \right) =$$

$$= \exp \left(-z \sum_{i=1}^{\infty} \log(1 - u y_i) \right) = \prod_{i=1}^{\infty} (1 - u y_i)^{-z}.$$

The desired formula then takes the form

$$\prod_{i=1}^{\infty} (1 - u y_i)^{-z} = \sum_{\lambda \in \mathbb{Y}} \prod_{\square \in \lambda} \frac{z + c(\square)}{h(\square)} \cdot s_\lambda(u y_1, u y_2, \ldots).$$

Replacing $u y_i$ with y_i, we arrive at the identity

$$\prod_{i=1}^{\infty} (1 - y_i)^{-z} = \sum_{\lambda \in \mathbb{Y}} \prod_{\square \in \lambda} \frac{z + c(\square)}{h(\square)} \cdot s_\lambda(y_1, y_2, \ldots).$$

The coefficients of the Schur functions in the right-hand side are the polynomials in z, hence it suffices to prove this identity for $z = N = 1, 2, \ldots$.

It is well known (see, e.g., [72, Chapter I, Section 3, Example 4]) that

$$\prod_{\square \in \lambda} \frac{N + c(\square)}{h(\square)} = s_\lambda(\underbrace{1, \ldots, 1}_{N})$$

(this is the dimension of the irreducible representation of the group $GL(N, \mathbb{C})$ with the highest weight $(\lambda_1, \ldots, \lambda_N)$ if $\lambda_{N+1} = \lambda_{N+2} = \ldots = 0$, and 0 otherwise). Our identity takes the form

$$\prod_{i=1}^{\infty} (1 - y_i)^{-N} = \sum_{\lambda \in \mathbb{Y}} s_\lambda(\underbrace{1, \ldots, 1}_{N})\, s_\lambda(y_1, y_2, \ldots).$$

But this is a special case of Cauchy's identity; see Proposition 2.12. This completes the proof of the lemma. □

Proposition 11.10 *For every $z \in \mathbb{C} \setminus \mathbb{Z}$, the distinguished vector ξ_0 of the representation T_z is a cyclic vector.*

Proof Once again we use the fact that T_z is an inductive limit of the representations Reg^n. It implies that it is enough to show that ξ_0 is cyclic in Reg^n for every $n = 1, 2, \dots$. The biregular representation Reg^n of $G(n) = S(n) \times S(n)$ is equivalent to $\bigoplus_{\lambda \in \mathbb{Y}_n} \pi_\lambda \otimes \pi_\lambda$, where π_λ denotes the irreducible representation of $S(n)$ corresponding to λ.[1] A vector of Reg^n is $G(n)$-cyclic if and only if its projections on all components $\pi_\lambda \otimes \pi_\lambda$ are nonzero, because the representations $\pi_\lambda \otimes \pi_\lambda$ of the group $G(n)$ are irreducible and pairwise distinct. Therefore, to prove that ξ_0 is cyclic in Reg^n it suffices to check that all coefficients in the expansion of ξ_0 in the basis $\{\chi^\lambda\}$ are nonzero, i.e., all numbers $M_z^{(n)}(\lambda)$ are nonzero. For $z \notin \mathbb{Z}$ this obviously follows from (11.2). □

Denote by φ_z the matrix element of T_z corresponding to the distinguished vector. Recall that the corresponding character of $S(\infty)$ is denoted by χ_z. It was described in Proposition 11.8 in terms of the corresponding coherent system of distributions.

If $z \notin \mathbb{Z}$ then Proposition 11.10 says that T_z is a spherical type representation in the sense of Definition 9.1. Therefore, in this case, all the information about representation T_z is contained in the spherical function φ_z.

Corollary 11.11 *For every $z \in \mathbb{C}$, the representations T_z and $T_{\bar{z}}$ are equivalent.*

Proof The key observation is that formula (11.2) is obviously invariant under conjugation $z \to \bar{z}$. It follows that $\chi_z = \chi_{\bar{z}}$ and consequently $\varphi_z = \varphi_{\bar{z}}$.

If $z \in \mathbb{R}$ then $\bar{z} = z$ and there is nothing to prove, so that we may assume $z \in \mathbb{C} \setminus \mathbb{R}$. Then, by virtue of Proposition 11.10, the distinguished vectors of both representations are cyclic. As we have just pointed out, their matrix coefficients coincide. Therefore, the representations are equivalent. □

11.1 Exercises

Exercise 11.1 Find all maps $p : S(n+1) \to S(n)$ that commute with the two-sided action of $S(n)$ for $n = 2$ and 3.

[1] For a finite group G the biregular representation decomposes as $\oplus_{\pi \in \hat{G}} \pi \otimes \bar{\pi}$, but for the symmetric group $\overline{\pi_\lambda} = \pi_\lambda$ (the symmetric group characters are real-valued because this group is *ambivalent*, i.e., every group element and its inverse lie in the same conjugacy class).

Exercise 11.2 Using the definition of the projective limit topology, prove that the space of virtual permutations $\mathfrak{S}$ is compact.

Exercise 11.3 Prove that if the homeomorphism of Proposition 11.2 takes $\{x_n\} \in \mathfrak{S}$ to $\{i_n\} \in I$ then $x_n = (n, i_n)(n-1, i_{n-1}) \cdots (1, i_1)$.

Exercise 11.4 The fact that the measure μ_t^n (see the definition just before Proposition 11.5) is a probability measure is equivalent to the identity

$$\sum_{k=0}^{n} c(n,k)t^k = t(t+1)\cdots(t+n-1),$$

where $c(n,k)$ denotes the number of permutations $\sigma \in S(n)$ with k cycles, called the *signless Stirling number of the first kind.* Prove this identity directly (or find two proofs in Stanley [112, Proposition 1.3.4]).

Exercise 11.5 The Ewens measures μ_t, $t > 0$, and their limits μ_0 and μ_∞ obtained by limit transitions $t \to 0$ and $t \to \infty$, are precisely those K-invariant probability measures on $\mathfrak{S}$ that are also product measures under the identification $\mathfrak{S}$ and the product space I of Proposition 11.2.

Exercise 11.6 The Ewens measures $\mu_t, 0 \le t \le \infty$, are mutually singular.

Exercise 11.7 (a) Show that for every $n = 1, 2, \ldots$, the isometry $L_z^n \colon H^n \to H^{n+1}$ admits a continuous continuation, with respect to the parameter $z \in \mathbb{C}^*$, to the points $z = 0$ and $z = \infty$ of the Riemann sphere $\mathbb{C} \cup \{\infty\}$. Conclude that the definition of the inductive limit representation of Proposition 11.7 also makes sense for the values $z = 0$ and $z = \infty$.

(b) Prove that the representation thus obtained for $z = \infty$ is equivalent to the natural two-sided regular representation of the group $G = S(\infty) \times S(\infty)$ on the Hilbert space $\ell^2(S(\infty))$.

Exercise 11.8 Denote by sgn the (one-dimensional) sign representation of $S(\infty)$. Show that for every $z \in \mathbb{C}$, T_{-z} is equivalent to $T_z \otimes (\mathrm{sgn} \times \mathrm{sgn})$.

Exercise 11.9 Prove directly that in accordance with Exercise 11.7, item (b),

$$\lim_{z\to\infty} M_z^{(n)}(\lambda) = \frac{\dim^2 \lambda}{n!}, \qquad \lambda \in \mathbb{Y}_n.$$

11.2 Notes

Our exposition follows the paper [67] by Kerov, Olshanski, and Vershik (its announcement was published much earlier as [66]).

12

Disjointness of Representations T_z

12.1 Preliminaries

Recall (Definition 9.1) that two unitary representations T_1 and T_2 of a group are said to be disjoint if T_1 and T_2 do not have equivalent nontrivial subrepresentations. Equivalently, there are no nonzero intertwining operators between T_1 and T_2, that is, operators $H(T_1) \to H(T_2)$ commuting with the action of the group.

Our aim is to prove the following result:

Theorem 12.1 *Assume that parameter z ranges over the set $\{z \in \mathbb{C} : \Im z \geq 0,\ z \notin \mathbb{Z}\}$. Then the representations T_z corresponding to distinct values of z are pairwise disjoint.*

Comments. (a) The assumption $\Im z \geq 0$ is introduced here because $T_z \sim T_{\bar{z}}$ (Corollary 11.11).

(b) The assumption $z \notin \mathbb{Z}$ can be dropped, see [67].

(c) The assertion of the theorem is important because it shows that the parameter z is a substantial element of the construction. This is not at all evident, especially because the biregular representations Reg^n from which every T_z is built do not depend on z; only the embeddings $\operatorname{Reg}^n \to \operatorname{Reg}^{n+1}$ do. So, *a priori* one could imagine that there are some intertwining operators between representations T_z with distinct values of parameter z.

Denote by P_z the spectral measure of the spherical function φ_z. Applying Theorem 9.2, we reduce Theorem 12.1 to the following assertion:

Theorem 12.2 *Assume that parameter z ranges over the set $\{z \in \mathbb{C} : \Im z \geq 0,\ z \notin \mathbb{Z}\}$. Then the measures P_z corresponding to distinct values of z are pairwise disjoint.*

The proof is given in the final section.

We already know that P_z is the boundary measure for the coherent system $\{M_z^{(n)}\}$ described in Proposition 11.8.

Our argument relies on the possibility to interpret coherent systems as Gibbs measures on the path space of the Young graph; see Chapter 7. Let us now describe this interpretation in more detail.

Recall that a (monotone) *path* in the Young graph $\mathbb{Y}$ is a (finite or infinite) sequence of vertices

$$\tau = (\tau_k \nearrow \tau_{k+1} \nearrow \cdots), \qquad \tau_i \in \mathbb{Y}_i.$$

Let $\mathscr{T}$ be the set of all infinite paths starting at $\varnothing$. This is a subset of the infinite product set $\prod_{n=0}^{\infty} \mathbb{Y}_n$. We endow $\mathscr{T}$ with the induced topology. Since $\prod_{n=0}^{\infty} \mathbb{Y}_n$ is a compact space and $\mathscr{T}$ is a closed subset, it is a compact space, too.

Given a finite path starting at $\varnothing$, $\sigma = (\sigma_0 = \varnothing \nearrow \sigma_1 \nearrow \cdots \nearrow \sigma_n)$, we denote by $C(\sigma)$ the cylinder subset in $\mathscr{T}$ formed by all paths $\tau \in \mathscr{T}$ coinciding with σ up to level n: $\tau_i = \sigma_i$ for $0 \le i \le n$. Notice that $C(\sigma)$ is an open and closed subset of $\mathscr{T}$. Any probability Borel measure on $\mathscr{T}$ is uniquely determined by its values on the subsets of the form $C(\sigma)$.

According to Definition 7.10, a probability measure on the path space $\mathscr{T}$ is said to be a *Gibbs measure* if its value on an arbitrary cylinder subset of the form $C(\sigma)$ depends on the endpoint of σ only.

Recall also (Proposition 7.11) that there exists a 1–1 correspondence $\{M^{(n)}\} \leftrightarrow \widetilde{M}$ between coherent systems of distributions on the Young graph $\mathbb{Y}$ and probability Gibbs measures on the path space $\mathscr{T}$. The correspondence is determined in the following way: for any cylinder set of the form $C(\sigma)$,

$$\widetilde{M}(C(\sigma)) = M^{(n)}(\lambda)/\dim\lambda, \qquad \lambda := \text{endpoint of } \sigma, \quad n = |\lambda|.$$

Equivalently,

$$\widetilde{M}(C(\sigma)) = \psi(\lambda), \qquad \lambda := \text{endpoint of } \sigma, \quad n = |\lambda|,$$

where ψ stands for the harmonic function on the vertices of $\mathbb{Y}$, associated with $\{M^{(n)}\}$ (see Definition 3.6).

Given a probability measure on the path space $\mathscr{T}$, we may speak about *random* infinite paths. Then the above relation says that $\psi(\lambda)$ equals the probability that the $\widetilde{M}$-random path goes along a fixed finite path joining $\varnothing$ with λ.

Note that there is a useful characterization of Gibbs measures as invariant measures with respect to a countable group of transformations of $\mathscr{T}$. This group is defined as follows. First, for each n we let $\mathscr{G}(n)$ be the group of the transformations $g : \mathscr{T} \to \mathscr{T}$ such that for any path $\tau = (\tau_n) \in \mathscr{T}$, we have $\tau_m = (g(\tau))_m$ for all $m \geq n$. Clearly, this is a finite group and we have $\mathscr{G}(n) \subset \mathscr{G}(n+1)$. Next, we define the group $\mathscr{G}$ as the union of the groups $\mathscr{G}(n)$.

Proposition 12.3 *A measure on $\mathscr{T}$ is Gibbs if and only if it is invariant under the action of $\mathscr{G}$.*

The proof is an easy exercise.

Define the *support* of a coherent system $M = \{M^{(n)}\}$ as the subset

$$\operatorname{supp}(M) = \{\lambda \in \mathbb{Y} : M^{|\lambda|}(\lambda) \neq 0\} \subset \mathbb{Y}.$$

The measure $\widetilde{M}$ is concentrated on the subspace of paths entirely contained in $\operatorname{supp}(M)$. We may view $\widetilde{M}$ as the law of a *Markov growth process* of Young diagrams, with the state set $\operatorname{supp}(M)$, discrete time $n = 0, 1, 2, \ldots$, and the *transition probabilities*

$$p(\lambda, \nu) = \operatorname{Prob}\{\tau_{n+1} = \nu \mid \tau_n = \lambda\}, \qquad \lambda \in \mathbb{Y}_n, \quad \nu \in \mathbb{Y}_{n+1},$$

where $\tau = (\tau_n)$ is the random path. The transition probabilities $p(\lambda, \nu)$ are unambiguously defined for all $\lambda \in \operatorname{supp}(M)$ by

$$p(\lambda, \nu) = \frac{M^{(n+1)}(\nu)}{M^{(n)}(\lambda)} \cdot \frac{\dim \lambda}{\dim \nu}, \qquad \lambda \in \operatorname{supp}(M) \cap \mathbb{Y}_n.$$

The system of transition probabilities uniquely determines the initial Gibbs measure, so that distinct Gibbs measures have distinct transition probabilities. On the other hand, all Gibbs measures have one and the same system of *cotransition probabilities*, which are nothing but the quantities $\Lambda^n_{n-1}(\lambda, \mu)$ introduced in (7.4). That is, we have

$$\Lambda^n_{n-1}(\lambda, \mu) = \operatorname{Prob}\{\tau_{n-1} = \mu \mid \tau_n = \lambda\}, \qquad \mu \in \mathbb{Y}_{n-1}, \quad \lambda \in \mathbb{Y}_n.$$

12.2 Reduction to Gibbs Measures

Since the boundary of the Young graph is the Thoma simplex Ω, Corollary 7.12 tells us that there is a one-to-one correspondence $P \leftrightarrow \widetilde{M}$ between probability measures P on Ω and probability Gibbs measures $\widetilde{M}$ on the path space $\mathscr{T}$.

Proposition 12.4 *Let P_1 and P_2 be two probability measures on Ω, and let $\widetilde{M}_1$ and $\widetilde{M}_2$ be the corresponding Gibbs measures on $\mathscr{T}$. Then P_1 and P_2 are disjoint if and only if $\widetilde{M}_1$ and $\widetilde{M}_2$ are disjoint.*

Proof First, introduce a notation. Given two finite (not necessarily normalized) measures ν_1, ν_2 on measurable space, let us denote by $\nu_1 \wedge \nu_2$ their greatest lower bound. Its existence can be verified as follows. Let f_1 and f_2

be the Radon–Nikodým derivatives of ν_1 and ν_2 with respect to $\nu_1+\nu_2$, then we set $\nu_1 \wedge \nu_2 = \min(f_1, f_2)(\nu_1+\nu_2)$. Observe that ν_1 and ν_2 are disjoint if and only if $\nu_1 \wedge \nu_2 = 0$.

Next, observe that the correspondence $P \leftrightarrow \widetilde{M}$ can be extended to finite, not necessarily normalized, measures.

Now we can proceed to the proof. In one direction the implication is trivial. Namely, if P_1 and P_2 are not disjoint, then $P_1 \wedge P_2$ is a nonzero measure. Let $\widetilde{M}$ be the corresponding Gibbs measure; it is nonzero because so is $P_1 \wedge P_2$. From the integral representation of coherent systems (Theorem 3.12) it follows that $\widetilde{M} \le \widetilde{M}_1$ and $\widetilde{M} \le \widetilde{M}_2$, so that $\widetilde{M}_1 \wedge \widetilde{M}_2 \ne 0$, whence $\widetilde{M}_1$ and $\widetilde{M}_2$ are not disjoint.

In the opposite direction, assume that $\widetilde{M}_1$ and $\widetilde{M}_2$ are not disjoint, so that $\widetilde{M}_1 \wedge \widetilde{M}_2$ is nonzero. We claim that $\widetilde{M}_1 \wedge \widetilde{M}_2$ is a Gibbs measure. Indeed, this follows from the characterization of Gibbs measures as invariant measures with respect to a countable group action, as explained in Proposition 12.3. Now, let P be the measure on Ω corresponding to $\widetilde{M}_1 \wedge \widetilde{M}_2$. It is a nonzero measure. Next, since $\widetilde{M}_1 \wedge \widetilde{M}_2 \le \widetilde{M}_1$ and $\widetilde{M}_1 \wedge \widetilde{M}_2 \le \widetilde{M}_2$, we also have $P \le P_1$, $P \le P_2$. (Indeed, this claim can be restated as follows: if $\widetilde{M}$, $\widetilde{M}'$ are two Gibbs probability measures such that $\widetilde{M} \le \text{const}\, \widetilde{M}'$ then the same inequality holds for the corresponding spectral measures on Ω, and the latter claim follows from the Approximation Theorem 6.14.) Therefore, P_1 and P_2 are not disjoint. $\square$

12.3 Exclusion of Degenerate Paths

Let p, q be two nonnegative integers, not equal to 0 simultaneously. The *fat hook* with parameters (p, q) is the set

$$\Gamma(p,q) := \{(i,j) \mid 1 \le i \le p,\ j = 1,2,\dots\} \cup \{(i,j) \mid 1 \le j \le q,\ i = 1,2,\dots\}.$$

(Note that $\Gamma(p,0)$ is actually not a hook but the horizontal strip of width p. Likewise, $\Gamma(0,q)$ is the vertical strip of width q.)

Denote by $\mathscr{T}(p,q)$ the set of paths $\tau = (\tau_n) \in \mathscr{T}$ such that $\tau_n \subset \Gamma(p,q)$ for all n. (Equivalently, τ_n does not contain the box $(p+1, q+1)$.) Let us say that a path $\tau \in \mathscr{T}$ is *degenerate* if it is contained in some set $\mathscr{T}(p,q)$.

We are going to prove the following proposition:

Proposition 12.5 *Assume $z \in \mathbb{C} \setminus \mathbb{Z}$ and let $\widetilde{M}_z$ be the Gibbs measure on $\mathscr{T}$ corresponding to the spectral measure P_z. The set of degenerate paths is a null set with respect to measure $\widetilde{M}_z$.*

Proof *Step 1.* Let $p_z(\lambda, \nu)$ be the transition probabilities of $\widetilde{M}_z$. We claim that for any fixed $p, q \in \mathbb{Z}_{\geq 0}$ with $p+q > 0$ there exists $\varepsilon > 0$ depending on z, p, q only, with the following property. If $\lambda \subset \Gamma(p,q)$ is an arbitrary Young diagram such that the set $\nu = \lambda \cup \{(p+1, q+1)\}$ is also a diagram (this is equivalent to saying that λ contains the boxes $(p+1, q)$ and $(p, q+1)$ but not the box $(p+1, q+1)$). Then

$$p_z(\lambda, \nu) \geq \varepsilon / n, \qquad n = |\lambda|.$$

Indeed, since the content of the box ν/λ is $q - p$, we have

$$p_z(\lambda, \nu) = \frac{|z+q-p|^2}{|z|^2+n} \cdot \frac{\dim \nu}{(n+1)\dim \lambda}.$$

The first factor can be easily estimated from below: since $z \notin \mathbb{Z}$, there exists $\varepsilon_1 > 0$ depending on z only, such that

$$\frac{|z+q-p|^2}{|z|^2+n} \geq \frac{\varepsilon_1}{n}.$$

Now consider the second factor. It follows from the hook formula that

$$\frac{\dim \nu}{(n+1)\dim \lambda} = \prod_b \frac{h(b)}{h(b)+1},$$

where the product is taken over the boxes $b \in \lambda$ such that either the arm or the leg of b (with respect to ν) contains the box $(p+1, q+1)$, and $h(b)$ denotes the hook-length of b in λ. There are exactly $p+q$ such boxes b, namely

$$(p+1, j), \quad 1 \leq j \leq q; \qquad (i, q+1), \quad 1 \leq i \leq p.$$

Therefore, there is a product of $p+q$ factors of the form $k/(k+1)$, where $k = 1, 2, \ldots$. Each of the factors is greater or equal to $1/2$, and the entire product is not less than $2^{-(p+q)}$. This provides the required estimate.

Step 2. Let us prove that $\widetilde{M}_z(\mathscr{T}(p,q)) = 0$ for every fixed $p, q \in \mathbb{Z}_{\geq 0}$ with $p+q > 0$. Denote by $\mathscr{T}'(p,q)$ the set of those paths $\tau \in \mathscr{T}(p,q)$ that are not contained in the smaller set $\mathscr{T}(p-1,q) \cup \mathscr{T}(p, q-1)$. It suffices to prove that $\mathscr{T}'(p,q)$ has measure 0 with respect to $\widetilde{M}_z$.

Let μ be an arbitrary diagram in $\Gamma(p,q)$ that contains the boxes $(p+1, q)$ and $(p, q+1)$, set $m = |\mu|$, and denote by $\mathscr{T}(p, q; \mu)$ the set of paths $\tau \in \mathscr{T}(p,q)$ with $\tau_m = \mu$. By the very definition of $\mathscr{T}'(p,q)$, for any path $\tau = (\tau_n) \in \mathscr{T}'(p,q)$ there exists a number n such that the diagram τ_n contains the boxes $(p+1, q)$ and $(p, q+1)$. Consequently the set $\mathscr{T}'(p,q)$ coincides with the union of the sets of the form $\mathscr{T}(p, q; \mu)$. Since there are countably many such sets, it remains to prove that each of them has measure 0.

It will be convenient to look at the measure $\widetilde{M}_z$ as describing a growth Markov process with the transition function $p_z(\lambda, \nu)$. Set

$$p_n = \text{Prob}\{\tau_{n+1} \subset \Gamma(p, q) \mid \mu \subseteq \tau_n \subset \Gamma(p, q)\}.$$

The measure of the set $\mathscr{T}(p, q; \mu)$ coincides with the probability of the event $\tau_m = \mu$, multiplied by the product of the conditional probabilities $\prod_{n \geq m} p_n$.

By virtue of step 1, we have

$$p_n \leq 1 - \frac{\varepsilon}{n}$$

so that $\prod_{n \geq m} p_n = 0$. □

12.4 Proof of Disjointness

Fix two distinct numbers z_1, z_2 in the upper half-plane $\Im z \geq 0$, which are not integers. We will prove that the spectral measures P_{z_1} and P_{z_2} are disjoint – this is the claim of Theorem 12.2.

By virtue of Proposition 12.4, it suffices to prove that the corresponding Gibbs measures $\widetilde{M}_{z_1}$ and $\widetilde{M}_{z_2}$ are disjoint. To simplify the notation, we set $\widetilde{M}_1 = \widetilde{M}_{z_1}$, $\widetilde{M}_2 = \widetilde{M}_{z_2}$. We also denote by $\{M_1^{(n)}\}$ and $\{M_2^{(n)}\}$ the corresponding coherent systems.

Recall that if $z \in \mathbb{C} \setminus \mathbb{Z}$, then the measure $M_z^{(n)}$ has nonzero weights $M_z^{(n)}(\lambda)$ for all $\lambda \in \mathbb{Y}_n$, and we have an explicit formula for $M_z^{(n)}(\lambda)$, see (11.2). Our arguments substantially rely on this formula.

Introduce a sequence $g_n(\tau)$ of functions on $\mathscr{T}$,

$$g_n(\tau) = \frac{M_2^{(n)}(\tau_n)}{M_1^{(n)}(\tau_n)}, \qquad n = 1, 2, \ldots, \qquad \tau = (\tau_n) \in \mathscr{T}.$$

Let X be the set of paths $\tau \in \mathscr{T}$ such that the sequence $(g_n(\tau))_{n \geq 1}$ converges, as $n \to \infty$, to a finite nonzero limit. This is a Borel subset of $\mathscr{T}$.

Lemma 12.6 *We have* $\widetilde{M}_1(X) = \widetilde{M}_2(X) = 0$.

Proof We shall show that X is contained in the union of the sets $\mathscr{T}(p, q)$, so that the claim will follow from Proposition 12.5.

Denote by $c_k(\tau)$ the content of the kth box $\tau_k \setminus \tau_{k-1}$. From (11.2) we get

$$g_n(\tau) = \prod_{k=1}^{n} \left| \frac{z_2 + c_k(\tau)}{z_1 + c_k(\tau)} \right|^2 \frac{|z_1|^2 + k - 1}{|z_2|^2 + k - 1}.$$

Therefore, X consists of those paths τ for which the infinite product

$$g_\infty(\tau) = \prod_{k=1}^{\infty} \left| \frac{z_2 + c_k(\tau)}{z_1 + c_k(\tau)} \right|^2 \frac{|z_1|^2 + k - 1}{|z_2|^2 + k - 1}$$

converges. In particular, the kth factor in the product should go to 1. Since the second fraction in right-hand side converges to 1, as $k \to \infty$, we conclude that

$$\lim_{k\to\infty} \left| \frac{z_2 + c_k(\tau)}{z_1 + c_k(\tau)} \right|^2 = 1, \qquad \tau \in X.$$

It follows from our assumptions on z_1, z_2 that the equality $|z_2 + c|^2 = |z_1 + c|^2$ may hold for at most one real number c. Indeed, this equation on c describes the set of points c that are equidistant from $-z_1$ and $-z_2$. Since $z_1 \neq z_2$, this set is a line in the complex plane $\mathbb{C}$, which cannot coincide with the real axis $\mathbb{R}$, because z_1, z_2 are both in the upper half-plane. Thus, the line is either parallel to $\mathbb{R}$ (then there is no real c at all) or intersects $\mathbb{R}$ at a single point.

Now, we fix an arbitrary integer c such that

$$\left| \frac{z_2 + c}{z_1 + c} \right|^2 \neq 1.$$

For any $\tau \in X$, the existence of the limit above implies that there is only a finite number of integers k such that $c_k(\tau) = c$. This means that any path $\tau \in X$ may contain only a finite number of boxes (p, q) on the diagonal $q - p = c$. Therefore, τ is contained in some subset of type $\mathscr{T}(p, q)$, which completes the proof. □

Lemma 12.7 *Let $\widetilde{A}$ and $\widetilde{B}$ be two Gibbs probability measures on $\mathscr{T}$ and $\{A^{(n)}\}$, $\{B^{(n)}\}$ be the corresponding coherent systems. Assume $\widetilde{A} \leq \operatorname{const} \widetilde{B}$ and let $f(\tau)$ denote the Radon–Nikodým derivative of $\widetilde{A}$ with respect to $\widetilde{B}$. Assume further that $B^{(n)}(\lambda) \neq 0$ for all n and all $\lambda \in \mathbb{Y}_n$. Then*

$$\lim_{n\to\infty} \frac{A^{(n)}(\tau_n)}{B^{(n)}(\tau_n)} = f(\tau)$$

for almost all paths $\tau = (\tau_n) \in \mathscr{T}$ with respect to $\widetilde{B}$.

Proof Let $\mathscr{T}^{[n]}$ denote the set of finite paths in $\mathbb{Y}$ going from $\varnothing$ to a vertex in $\mathbb{Y}_n$. There is a natural projection $\mathscr{T} \to \mathscr{T}^{[n]}$ assigning to a path τ its finite part $\tau^{[n]} = (\tau_0, \dots, \tau_n)$. Notice that the infinite path space $\mathscr{T}$ can be identified with the projective limit space $\varprojlim \mathscr{T}^{[n]}$.

Denote by $\Sigma^{[n]}$ the finite algebra of cylinder subsets with the bases in $\mathscr{T}^{[n]}$. The algebras $\Sigma^{[n]}$, $n = 1, 2, \ldots$, form an increasing family generating the sigma-algebra Σ of Borel sets in $\mathscr{T}$.

Consider the probability space $(\mathscr{T}, \Sigma, \widetilde{B})$. The function f is bounded and $\widetilde{B}$-measurable. Hence, by the martingale theorem (cf., e.g., Shiryaev [110, Ch. VII, Section 4, Theorem 3]),

$$\lim_{n\to\infty} \mathbb{E}(f \mid \Sigma^{[n]}) = f$$

almost everywhere.

On the other hand, let $A^{[n]}$ and $B^{[n]}$ be the pushforwards of the measures $\widetilde{A}$ and $\widetilde{B}$ taken with respect to the projection $\mathscr{T} \to \mathscr{T}^{[n]}$. The conditional expectation $\mathbb{E}(f \mid \Sigma^{[n]})$ is nothing but the function

$$f_n(\tau) = \frac{A^{[n]}(\tau^{[n]})}{B^{[n]}(\tau^{[n]})}.$$

Since $\widetilde{A}$ is a Gibbs measure, we have

$$A^{[n]}(\tau^{[n]}) = \frac{1}{\dim \tau_n} A^{(n)}(\tau_n),$$

and similarly

$$B^{[n]}(\tau^{[n]}) = \frac{1}{\dim \tau_n} B^{(n)}(\tau_n).$$

It follows that

$$f_n(\tau) = \frac{A^{(n)}(\tau_n)}{B^{(n)}(\tau_n)},$$

and the proof is completed. □

Now we are in a position to show that the measures $\widetilde{M}_1$ and $\widetilde{M}_2$ are disjoint. Set $\widetilde{A} = \widetilde{M}_1$, $\widetilde{B} = (\widetilde{M}_1 + \widetilde{M}_2)/2$. Then $\widetilde{A} \le 2\widetilde{B}$ and hence the Radon–Nikodým derivative of $\widetilde{A}$ with respect to $\widetilde{B}$ is well defined. Denote it by $f(\tau)$. We have $0 \le f(\tau) \le 2$. The measures $\widetilde{M}_1$ and $\widetilde{M}_2$ are disjoint if and only if $f(\tau)$ takes only two values 0 and 2, almost surely with respect to the measure $\widetilde{B}$.

On the other hand, by virtue of Lemma 12.7, $f(\tau)$ is $\widetilde{B}$-almost surely the limit of the functions $f_n(\tau)$. Let Y be the set of those paths τ for which the limit of $f_n(\tau)$ exists and is distinct from 0 and 2. Observe that

$$\begin{aligned} f_n(\tau) &= \frac{A^{(n)}(\tau_n)}{B^{(n)}(\tau_n)} = 2\frac{M_1^{(n)}(\tau_n)}{M_1^{(n)}(\tau_n) + M_2^{(n)}(\tau_n)} \\ &= 2\left(1 + \frac{M_2^{(n)}(\tau_n)}{M_1^{(n)}(\tau_n)}\right)^{-1} = \frac{2}{1 + g_n(\tau)}. \end{aligned}$$

Consequently, Y coincides with the set of those paths τ for which $g_n(\tau)$ has a finite nonzero limit, that is, $Y = X$. But $\widetilde{M}_1(X) = \widetilde{M}_2(X) = 0$ by virtue of Lemma 12.6. Hence, $\widetilde{B}(Y) = \widetilde{B}(X) = 0$, so that $f(\tau)$ is 0 or 2 almost surely with respect to $\widetilde{B}$.

This completes the proof. □

12.5 Exercises

Exercise 12.1 Let us say that a path $\tau = (\tau_n) \in \mathscr{T}$ is *regular* if the sequence of its vertices τ_n converges to a point of Ω in the sense of Definition 6.15. That is, $\frac{1}{n}\omega_{\tau_n} \to \omega \in \Omega$ or, equivalently, the limits (6.18) exist.

Prove that the regular paths form a Borel subset in $\mathscr{T}$.

Prove that any Gibbs measure on $\mathscr{T}$ is concentrated on the subset of regular paths.

12.6 Notes

Here again we closely follow Kerov–Olshanski–Vershik [67]. As we pointed out at the end of the Introduction, the spectral decomposition of representations T_z was described in Borodin–Olshanski [9] (for $z \in \mathbb{C} \setminus \mathbb{Z}$ – the nondegenerate case) and in [67] (for $z \in \mathbb{Z}$ – the degenerate case).

References

[1] M. Aissen, A. Edrei, I.J. Schoenberg, and A. Whitney, On the generating functions of totally positive sequences. *Proc. Nat. Acad. Sci. USA* 37 (1951), 303–307.

[2] M. Aissen, I.J. Schoenberg, and A. Whitney, On the generating functions of totally positive sequences I. *J. Analyse Math.* 2 (1952), 93–103.

[3] D.J. Aldous, Exchangeability and related topics. In: *École d'Été de Probabilités de Saint-Flour, XIII–1983*, Springer Lecture Notes in Math. 1117, Springer, 1985, pp. 1–198.

[4] J. Baik, P. Deift, and K. Johansson, On the distribution of the length of the longest increasing subsequence of random permutations. *J. Amer. Math. Soc.* 12 (1999), 1119–1178; arXiv:math/9810105.

[5] V.I. Bogachev, *Measure Theory*. Springer, 2007.

[6] S. Bochner, *Harmonic Analysis and the Theory of Probability*. University of California Press, 1955.

[7] A. Borodin, A. Okounkov, and G. Olshanski, Asymptotics of Plancherel measures for symmetric groups, *J. Amer. Math. Soc.* 13 (2000), no. 3, 481–515.

[8] A. Borodin and G. Olshanski, Point processes and the infinite symmetric group. *Mathematical Research Letters* 5 (1998), 799–816.

[9] A. Borodin and G. Olshanski, Distributions on partitions, point processes and the hypergeometric kernel. *Commun. Math. Phys.* 211 (2000), 335–358; arXiv:math/9904010.

[10] A. Borodin and G. Olshanski, Harmonic functions on multiplicative graphs and interpolation polynomials. *Electronic J. Comb.* 7 (2000), #R28; arXiv:math/9912124.

[11] A. Borodin and G. Olshanski, Infinite random matrices and ergodic measures. *Comm. Math. Phys.* 223 (2001), no. 1, 87–123; arXiv:math-ph/0010015.

[12] A. Borodin and G. Olshanski, Z-Measures on partitions, Robinson–Schensted–Knuth correspondence, and $\beta = 2$ ensembles. In: *Random Matrix Models and their Applications* (P.M. Bleher and A.R. Its, eds). MSRI Publications, vol. 40, Cambridge University Press, 2001, pp. 71–94.

[13] A. Borodin and G. Olshanski, Harmonic analysis on the infinite-dimensional unitary group and determinantal point processes. *Ann. Math.* 161 (2005), no. 3, 1319–1422.

[14] A. Borodin and G. Olshanski, Representation theory and random point processes. In: *European Congress of Mathematics (ECM), Stockholm, Sweden,*

June 27–July 2, 2004 A. Laptev (ed.). European Mathematical Society, 2005, pp. 73–94.

[15] A. Borodin and G. Olshanski, Random partitions and the Gamma kernel. *Adv. Math.* 194 (2005),141–202; arXiv:math-ph/0305043.

[16] A. Borodin and G. Olshanski, Z-measures on partitions and their scaling limits. *Europ. J. Comb.* 26 (2005), no. 6, 795–834.

[17] A. Borodin and G. Olshanski, Markov processes on partitions. *Probab. Theory Rel. Fields* 135 (2006), 84–152; arXiv:math-ph/0409075.

[18] A. Borodin and G. Olshanski, Meixner polynomials and random partitions. *Moscow Math. J.* 6 (2006), no. 4, 629–655.

[19] A. Borodin and G. Olshanski, Stochastic dynamics related to Plancherel measure on partitions. In: *Representation Theory, Dynamical Systems, and Asymptotic Combinatorics* (V. Kaimanovich and A. Lodkin, eds). Amer. Math. Soc. Translations, Series 2: Advances in the Mathematical Sciences, vol. 217, 2006, pp. 9–21; arXiv:math-ph/0402064.

[20] A. Borodin and G. Olshanski, Infinite-dimensional diffusions as limits of random walks on partitions. *Probab. Theory Rel. Fields* 144 (2009), 281–318; arXiv:0706.1034.

[21] A. Borodin and G. Olshanski, Markov processes on the path space of the Gelfand-Tsetlin graph and on its boundary. *J. Funct. Anal.* 263 (2012), 248–303; arXiv:1009.2029.

[22] A. Borodin and G. Olshanski, The boundary of the Gelfand–Tsetlin graph: A new approach. *Advances in Math.* 230 (2012), 1738–1779; arXiv:1109.1412.

[23] A. Borodin and G. Olshanski, The Young bouquet and its boundary. *Moscow Math. J.* 13 (2013), no. 2; arXiv:1110.4458.

[24] A. Borodin and G. Olshanski, Markov dynamics on the Thoma cone: a model of time-dependent determinantal processes with infinitely many particles. *Electron. J. Probab.* 18 (2013), no. 75, 1–43.

[25] A. Borodin, G. Olshanski, and E. Strahov, Giambelli compatible point processes. *Advances in Appl. Math.* 37 (2006), 209–248.

[26] A. Bufetov and V. Gorin, Stochastic monotonicity in Young graph and Thoma theorem. *International Mathematics Research Notices*, vol. 2015, 12920–12940; arXiv:1411.3307.

[27] O. Bratteli, Inductive limits of finite dimensional C^*-algebras. *Trans. Amer. Math. Soc.* 171 (1972), 195–234.

[28] T. Ceccherini-Silberstein, F. Scarabotti, and F. Tolli, *Harmonic Analysis on Finite Groups – Representation Theory, Gelfand Pairs and Markov Chains*. Cambridge Studies in Advanced Mathematics 108, Cambridge University Press, 2008.

[29] T. Ceccherini-Silberstein, F. Scarabotti, and F. Tolli, *Representation Theory of the Symmetric Group, the Okounkov–Vershik Approach, Character Formulas, and Partition Algebras*. Cambridge Studies in Advanced Mathematics 121, Cambridge University Press, 2010.

[30] C.W. Curtis, *Pioneers of Representation Theory: Frobenius, Burnside, Schur, and Brauer*. History of Mathematics vol. 15. American Mathematical Society; London Mathematical Society, 1999.

[31] P. Diaconis, *Group Representations in Probability and Statistics*. Institute of Mathematical Statistics, 1988.

[32] P. Diaconis, D. Freedman, Partial exchangeability and sufficiency. In: *Statistics: Applications and New Directions* (Calcutta, 1981), pp. 205–236, Indian Statist. Inst., 1984.

[33] J. Dixmier, C^**-Algebras*. North-Holland, 1977.

[34] E.B. Dynkin, Initial and final behavior of trajectories of Markov processes. *Uspehi Mat. Nauk* 26 (1971), no. 4, 153–172 (Russian); English translation: *Russian Math. Surveys* 26 (1971), no. 4, 165–185.

[35] E.B. Dynkin, Sufficient statistics and extreme points. *Ann. Probab.* 6 (1978), 705–730.

[36] A. Edrei, On the generating functions of totally positive sequences II. *J. Analyse Math.,* 2 (1952), 104–109.

[37] A. Edrei, On the generating function of a doubly infinite, totally positive sequence. *Trans. Amer. Math. Soc.* 74 (1953), 367–383.

[38] W.J. Ewens and S. Tavaré, The Ewens sampling formula. In: *Encyclopedia of Statistical Science*, Vol. 2 (S. Kotz, C.B. Read, and D.L. Banks, eds.), pp. 230–234, Wiley, 1998.

[39] W. Feller, *An Introduction to Probability Theory and its Applications*, vol. 2. Wiley, 1966, 1971.

[40] V. Féray, Combinatorial interpretation and positivity of Kerov's character polynomials, *Journal of Algebraic Combinatorics*, 29 (2009), 473–507.

[41] F.G. Frobenius, Über die Charaktere der symmetrischen Gruppe. *Sitz. Konig. Preuss. Akad. Wissen.* (1900), 516–534; *Gesammelte Abhandlungen III*, Springer-Verlag, 1968, pp. 148–166.

[42] W. Fulton and J. Harris, *Representation Theory. A First Course*. Springer, 1991.

[43] A. Gnedin, The representation of composition structures, *Ann. Prob.* 25 (1997), 1437–1450.

[44] A. Gnedin and S. Kerov, The Plancherel measure of the Young-Fibonacci graph. *Math. Proc. Cambridge Philos. Soc.* 129 (2000), no. 3, 433–446.

[45] A. Gnedin and G. Olshanski, Coherent permutations with descent statistic and the boundary problem for the graph of zigzag diagrams. *Internat. Math. Research Notices* 2006 (2006), Article ID 51968; arXiv:math/0508131.

[46] A. Gnedin and G. Olshanski, The boundary of Eulerian number triangle. *Moscow Math. J.* 6 (2006), no. 3, 461–475; arXiv:math/0602610.

[47] A. Gnedin and G. Olshanski, A q-analogue of de Finetti's theorem. *Electronic Journal of Combinatorics* 16 (2009), no. 1, paper #R78; arXiv:0905.0367.

[48] A. Gnedin and G. Olshanski, q-Exchangeability via quasi-invariance, *Ann. Prob.* 38 (2010), mo. 6, 2103–2135; arXiv:0907.3275.

[49] A. Gnedin and G. Olshanski, The two-sided infinite extension of the Mallows model for random permutations. *Advances in Applied Math.* 48 (2012), Issue 5, 615–639; arXiv:1103.1498.

[50] A. Gnedin and J. Pitman, Exchangeable Gibbs partitions and Stirling triangles. *Zap. Nauchn. Semin. POMI* 325, 83–102 (2005); reproduced in *J. Math. Sci., New York* 138 (2006), no. 3, 5674–5685; arXiv:math/0412494.

[51] F.M. Goodman and S.V. Kerov, The Martin boundary of the Young–Fibonacci lattice. *J. Algebraic Combin.* 11 (2000), no. 1, 17–48.

[52] R. Goodman and N.R. Wallach, *Symmetry, Representations, and Invariants.* Springer, 2009.

[53] V. Gorin, Disjointness of representations arising in the problem of harmonic analysis on an infinite-dimensional unitary group. *Funct. Anal. Appl.* 44 (2010), no. 2, 92–105; arXiv:0805.2660.

[54] V.E. Gorin, Disjointness of representations arising in the problem of harmonic analysis on an infinite-dimensional unitary group. *Funktsional. Anal. i Prilozhen.* 44 (2010), no. 2, 14–32 (Russian); English translation in *Funct. Anal. Appl.* 44 (2010), 92–105.

[55] V. Gorin, The q-Gelfand–Tsetlin graph, Gibbs measures and q-Toeplitz matrices. *Adv. Math.* 229 (2012), no. 1, 201–266.

[56] E. Hewitt and L.J. Savage, Symmetric measures on Cartesian products. *Trans. Amer. Math. Soc.* 80 (1955), 470–501.

[57] G. James and A. Kerber, *The Representation Theory of the Symmetric Group.* Addison-Wesley, 1981.

[58] K. Johansson, Discrete orthogonal polynomial ensembles and the Plancherel measure. *Ann. Math. (2)* 153 (2001) 259–296; arXiv:math/9906120.

[59] S. Kerov, Generalized Hall–Littlewood symmetric functions and orthogonal polynomials. In: *Advances in Soviet Math.* vol. 9. Amer. Math. Soc., 1992, pp. 67–94.

[60] S. Kerov, Gaussian limit for the Plancherel measure of the symmetric group. *C. R. Acad. Sci. Paris Sér. I Math.* 316 (1993), no. 4, 303–308.

[61] S. Kerov, A differential model for the growth of Young diagrams. *Proceedings of the St. Petersburg Mathematical Society*, Vol. IV, pp. 111–130, Amer. Math. Soc. Transl. Ser. 2, 188, Amer. Math. Soc., 1999.

[62] S.V. Kerov, Anisotropic Young diagrams and Jack symmetric functions. *Funct. Anal. Appl.* 34 (2000), No.1, 45–51; arXiv:math/9712267.

[63] S.V. Kerov, *Asymptotic Representation Theory of the Symmetric Group and its Applications in Analysis.* Translations of Mathematical Monographs, 219. American Mathematical Society, 2003.

[64] S. Kerov, A. Okounkov, and G. Olshanski, The boundary of Young graph with Jack edge multiplicities. *Intern. Mathematics Research Notices*, (1998), no. 4, 173–199; arXiv:q-alg/9703037.

[65] S. Kerov and G. Olshanski, Polynomial functions on the set of Young diagrams. *Comptes Rendus Acad. Sci. Paris, Ser. I*, 319 (1994), 121–126.

[66] S. Kerov, G. Olshanski, and A. Vershik, Harmonic analysis on the infinite symmetric group. A deformation of the regular representation. *Comptes Rendus Acad. Sci. Paris, Sér. I* 316 (1993), 773–778.

[67] S. Kerov, G. Olshanski, and A. Vershik, Harmonic analysis on the infinite symmetric group. *Invent. Math.* 158 (2004), 551–642; arXiv:math/0312270.

[68] S.V. Kerov, O.A. Orevkova, Random processes with common cotransition probabilities. *Zapiski Nauchnyh Seminarov LOMI* 184 (1990), 169–181 (Russian); English translation in *J. Math. Sci. (New York)* 68 (1994), no. 4, 516–525.

[69] S.V. Kerov and A.M. Vershik, Characters, factor representations and K-functor of the infinite symmetric group. In: *Operator Algebras and Group Representations, Vol. II (Neptun, 1980)*, Monographs and Studies in Mathematics, vol. 18, Pitman, 1984, pp. 23–32.

[70] J.F.C. Kingman, The representation of partition structures. *J. London Math. Soc.* 18 (1978), 374–380.

[71] A. Kolmogoroff, *Grundbegriffe der Wahrscheinlichkeitsrechnung*. Springer, 1933. Second English edition: A.N. Kolmogorov, *Foundations of the Theory of Probability*. Chelsea Publ. Comp., 1956.

[72] I.G. Macdonald, *Symmetric Functions and Hall Polynomials*. 2nd edition. Oxford University Press, 1995.

[73] Yu. I. Manin, *Gauge Field Theory and Complex Geometry*. Translated from the Russian by N. Koblitz and J. R. King, Springer-Verlag, 1988, x + 295 pp.

[74] P.-A. Meyer, *Probability and Potentials*. Blaisdell, 1966.

[75] Yu. A. Neretin, Hua-type integrals over unitary groups and over projective limits of unitary groups. *Duke Math. J.* 114 (2002), 239–266; arXiv:math-ph/0010014.

[76] Yu. A. Neretin, The subgroup $PSL(2, R)$ is spherical in the group of diffeomorphisms of the circle. *Funct. Anal. Appl.*, 50 (2016), 160–162; arXiv:1501.05820.

[77] J. von Neumann, On infinite direct products. *Comp. Math.* 6 (1938), 1–77.

[78] A. Okounkov, Thoma's theorem and representations of the infinite bisymmetric group. *Funct. Anal. Appl.* 28 (1994), no. 2, 100–107.

[79] A. Okounkov, On representations of the infinite symmetric group. *Zapiski Nauchnyh Seminarov POMI* 240 (1997), 166–228 (Russian); English translation: *J. Math. Sci. (New York)* 96 (1999), No. 5, 3550–3589.

[80] A. Okounkov and G. Olshanski, Shifted Schur functions. *Algebra i analiz* 9 (1997), no. 2, 73–146 (Russian); English version: *St. Petersburg Mathematical J.*, 9 (1998), 239–300; arXiv:q-alg/9605042.

[81] A. Okounkov and G. Olshanski, Asymptotics of Jack polynomials as the number of variables goes to infinity. *Intern. Math. Research Notices* 1998 (1998), no. 13, 641–682; arXiv:q-alg/9709011.

[82] A. Okounkov and G. Olshanski, Limits of BC-type orthogonal polynomials as the number of variables goes to infinity. In: *Jack, Hall–Littlewood and Macdonald Polynomials*. Contemp. Math., 417, pp. 281–318. Amer. Math. Soc., 2006.

[83] G. Olshanski, Unitary representations of the infinite-dimensional classical groups $U(p,\infty)$, $SO(p,\infty)$, $Sp(p,\infty)$ and the corresponding motion groups. *Soviet Math. Doklady* 19 (1978), 220–224.

[84] G. Olshanski, Unitary representations of infinite-dimensional pairs (G, K) and the formalism of R. Howe. *Dokl. Akad. Nauk SSSR* 269 (1983), 33–36 (Russian); English translation in *Soviet Math., Doklady.* 27 (1983), 290–294.

[85] G. Olshanski, Unitary representations of (G, K)-pairs connected with the infinite symmetric group $S(\infty)$. *Algebra i analiz* 1 (1989), no. 4, 178–209 (Russian); English translation: *Leningrad Math. J.* 1, no. 4 (1990), 983–1014.

[86] G. Olshanski, Method of holomorphic extensions in the representation theory of infinitedimensinal classical groups. *Funct. Anal. Appl.* 22, no. 4 (1989), 273–285.

[87] G. Olshanski, Unitary representations of infinite-dimensional pairs (G, K) and the formalism of R. Howe. In: *Representation of Lie Groups and Related Topics* (A.M. Vershik and D.P. Zhelobenko, eds), Adv. Stud. Contemp. Math. 7, Gordon and Breach, 1990, pp. 269–463.

[88] G. Olshanski, Representations of infinite-dimensional classical groups, limits of enveloping algebras, and Yangians. In: *Topics in Representation Theory* (A. A. Kirillov, ed.). Advances in Soviet Math., vol. 2. Amer. Math. Soc., 1991, pp. 1–66.

[89] G. Olshanski, On semigroups related to infinite-dimensional groups. In: *Topics in Representation Theory* (A. A. Kirillov, ed.). Advances in Soviet Math., vol. 2. Amer. Math. Soc., 1991, pp. 67–101.

[90] G. Olshanski, Probability measures on dual objects to compact symmetric spaces, and hypergeometric identities. *Funkts. Analiz i Prilozh.* 37 (2003), no. 4 (Russian); English translation in *Functional Analysis and its Applications* 37 (2003), 281–301.

[91] G. Olshanski, The problem of harmonic analysis on the infinite-dimensional unitary group. *J. Funct. Anal.* 205 (2003), 464–524; arXiv:math/0109193.

[92] G. Olshanski, An introduction to harmonic analysis on the infinite symmetric group. In: *Asymptotic Combinatorics with Applications to Mathematical Physics* (A. Vershik, ed). Springer Lecture Notes in Math. 1815, 2003, pp. 127–160.

[93] G. Olshanski, Difference operators and determinantal point processes. *Funct. Anal. Appl.* 42 (2008), no. 4, 317–329; arXiv:0810.3751.

[94] G. Olshanski, Anisotropic Young diagrams and infinite-dimensional diffusion processes with the Jack parameter. *Intern. Math. Research Notices* 2010 (2010), no. 6, 1102–1166; arXiv:0902.3395.

[95] G. Olshanski, Laguerre and Meixner symmetric functions, and infinite-dimensional diffusion processes. *Zapiski Nauchnyh Seminarov POMI* 378 (2010), 81–110; reproduced in *J. Math. Sci. (New York)* 174 (2011), no. 1, 41–57; arXiv:1009.2037.

[96] G. Olshanski, Laguerre and Meixner orthogonal bases in the algebra of symmetric functions. *Intern. Math. Research Notices* 2012 (2012), 3615–3679 arXiv:1103.5848.

[97] G. Olshanski, A. Regev, and A. Vershik, Frobenius–Schur functions: summary of results; arXiv:math/0003031.

[98] G. Olshanski, A. Regev, and A. Vershik, Frobenius-Schur functions. In: *Studies in Memory of Issai Schur* (A. Joseph, A. Melnikov, R. Rentschler, eds.). Progress in Mathematics 210, pp. 251–300. Birkhäuser, 2003; arXiv:math/0110077.

[99] A.A. Osinenko, Harmonic analysis on the infinite-dimensional unitary group. *Zapiski Nauchnyh Seminarov POMI* 390 (2011), 237–285 (Russian); English translation in *J. Math. Sci. (New York)* 181 (2012), no. 6, 886–913.

[100] K.R. Parthasarathy, *Probability Measures on Metric Spaces*. Academic Press, 1967.

[101] R.R. Phelps, *Lectures on Choquet's Theorems*. Van Nostrand, 1966.

[102] D. Pickrell, Measures on infinite-dimensional Grassmann manifold. *J. Func. Anal.* 70 (1987), 323–356.

[103] D. Pickrell, Separable representations for automorphism groups of infinite symmetric spaces. *J. Funct. Anal.* 90 (1990), 1–26.

[104] J. Pitman, *Combinatorial Stochastic Processes. Lectures from the 32nd Summer School on Probability Theory held in Saint-Flour, July 7–24, 2002*. Springer Lecture Notes in Mathematics 1875, 2006.

[105] M. Reed and B. Simon, *Methods of Modern Mathematical Physics. Vol. 1. Functional Analysis*. Academic Press, 1972.

[106] M. Reed and B. Simon, *Methods of Modern Mathematical Physics. Vol. 2. Fourier Analysis. Self-Adjointness*. Academic Press, 1975.

[107] B.E. Sagan, *The Symmetric Group. Representations, Combinatorial Algorithms, and Symmetric Functions*. Second edition, Springer, 2001.

[108] S.A. Sawyer, Martin boundaries and random walks. In: *Harmonic Functions on Trees and Buildings*, Adam Koranyi (ed). Contemporary Mathematics 206, Providence, RI: American Mathematical Society, 1997, pp. 17–44.

[109] I.J. Schoenberg, Some analytical aspects of the problem of smoothing. In: *Studies and Essays Presented to R. Courant on his 60th Birthday, January 8, 1948*, New York, NY: Interscience Publishers, Inc., 1948, pp. 351–370.

[110] A.N. Shiryaev, *Probability*, Graduate Texts in Mathematics vol. 95. Springer, 1996.

[111] B. Simon, *Representations of Finite and Compact Groups*. Amer. Math. Soc., 1996.

[112] R.P. Stanley, *Enumerative Combinatorics, vol. 1*. Cambridge University Press, 1997.

[113] R.P. Stanley, *Enumerative Combinatorics, vol. 2*. Cambridge University Press, 1999.

[114] E. Strahov, Generalized characters of the symmetric group. *Adv. Math.* 212 (2007), no. 1, 109–142; arXiv:math/0605029.

[115] E. Strahov, A differential model for the deformation of the Plancherel growth process. *Adv. Math.* 217 (2008), no. 6, 2625–2663.

[116] E. Strahov, The z-measures on partitions, Pfaffian point processes, and the matrix hypergeometric kernel. *Adv. Math.* 224 (2010), no. 1, 130–168.

[117] E. Thoma, Die unzerlegbaren, positive-definiten Klassenfunktionen der abzählbar unendlichen, symmetrischen Gruppe. *Math. Zeitschr.*, 85 (1964), 40–61.

[118] A.M. Vershik, Intrinsic metric on graded graphs, standardness, and invariant measures. *Zapiski Nauchn. Semin. POMI* 421 (2014), 58–67 (Russian). English translation: *J. Math. Sci. (New York)* 200 no. 6 (2014), 677–681.

[119] A.M. Vershik, The problem of describing central measures on the path spaces of graded graphs. *Funct. Anal. Appl.* 48 (2014), no. 4, 256–271.

[120] A.M. Vershik, Equipped graded graphs, projective limits of simplices, and their boundaries. *Zapiski Nauchn. Semin. POMI* 432 (2015), 83–104 (Russian). English translation: *J. Math. Sci. (New York)* 209 (2015), 860–873.

[121] A.M. Vershik and S.V. Kerov, Characters and factor representations of the infinite symmetric group. *Dokl. Akad. Nauk SSSR* 257 (1981), 1037–1040 (Russian); English translation in *Soviet Math., Doklady* 23 (1981), 389–392.

[122] A.M. Vershik and S.V. Kerov, Asymptotic theory of characters of the symmetric group. *Funct. Anal. Appl.* 15 (1981), no. 4, 246–255.

[123] A.M. Vershik and S.V. Kerov, Characters and factor representations of the infinite unitary group. *Doklady AN SSSR* 267 (1982), no. 2, 272–276 (Russian); English translation: *Soviet Math. Doklady* 26 (1982), 570–574.

[124] A.M. Vershik and S.V. Kerov, The characters of the infinite symmetric group and probability properties of the Robinson–Schensted–Knuth correspondence. *SIAM J. Alg. Disc. Meth.* 7 (1986), 116–123.

[125] A.M. Vershik and S.V. Kerov, Locally semisimple algebras. Combinatorial theory and the K_0-functor. *J. Soviet Math.* 38 (1987), 1701–1733.

[126] A.M. Vershik and S.V. Kerov, The Grothendieck group of infinite symmetric group and symmetric functions (with the elements of the theory of K_0-functor for AF-algebas). In: *Representations of Lie Groups and Related Topics*. Advances in Contemp. Math., vol. 7 (A.M. Vershik and D.P. Zhelobenko, editors). New York, NY; London: Gordon and Breach, 1990, pp. 39–117.

[127] A.V. Vershik and N.I. Nessonov, Stable representations of the infinite symmetric group. *Izvestiya: Mathematics* 79 (2015), no. 6, 93–124; English translation: *Izvestiya: Mathematics* 79 (2015), no. 6, 1184–1214.

[128] G. Viennot, Maximal chains of subwords and up-down sequences of permutations. *J. Comb. Theory. Ser. A* 34 (1983), 1–14.

[129] E. B. Vinberg, *Linear Representations of Groups*. Birkhäuser, 1989. (English translation of the original Russian edition: Nauka, Moscow, 1985.)

[130] D. Voiculescu, Sur les représentations factorielles finies de $U(\infty)$ et autres groupes semblables. *C. R. Acad. Sci., Paris, Sér. A* 279 (1974), 945–946.

[131] D. Voiculescu, Représentations factorielles de type II_1 de $U(\infty)$. *J. Math. Pures et Appl.* 55 (1976), 1–20.

[132] A.J. Wassermann, Automorphic actions of compact groups on operator algebras. Unpublished Ph.D. dissertation. Philadelphia, PA: University of Pennsylvania, 1981.

[133] H. Weyl, *The Classical Groups. Their Invariants and Representations*. Princeton University Press, 1939; 1997 (fifth edition).

[134] G. Winkler, *Choquet Order and Simplices. With Applications in Probabilistic Models*. Springer Lect. Notes Math. 1145, Springer, 1985.

[135] A.V. Zelevinsky, *Representations of Finite Classical Groups. A Hopf Algebra Approach*. Springer Lecture Notes in Math. 869, Springer, 1981.

[136] D.P. Zhelobenko, *Compact Lie Groups and their Representations*. Nauka, 1970 (Russian); English translation: Transl. Math. Monographs 40, Amer. Math. Soc., 1973.

Index